Andreas Kerren   John T. Stasko
Jean-Daniel Fekete   Chris North (Eds.)

Information Visualization

Human-Centered Issues and Perspectives

With 15 colored illustrations

 Springer

Andreas Kerren
Växjö University
School of Mathematics and Systems Engineering
Computer Science Department
Vejdes Plats 7, 351 95 Växjö, Sweden
E-mail: kerren@acm.org

John T. Stasko
Georgia Institute of Technology
School of Interactive Computing and GVU Center
85 5th St., NW, Atlanta, GA 30332-0760, USA
E-mail: stasko@cc.gatech.edu

Jean-Daniel Fekete
INRIA Saclay - Île-de-France Research Centre
Bat. 490, Université Paris-Sud
91405 Orsay Cedex, France
E-mail: jean-daniel.fekete@inria.fr

Chris North
Virginia Tech, Department of Computer Science
and Center for Human-Computer Interaction
2202 Kraft Drive, Blacksburg, VA 24061-0106, USA
E-mail: north@vt.edu

Library of Congress Control Number: 2008931479

CR Subject Classification (1998): H.2.8, H.5, I.2.4, I.3

LNCS Sublibrary: SL 3 – Information Systems and Application, incl. Internet/Web
and HCI

ISSN   0302-9743
ISBN-10   3-540-70955-X Springer Berlin Heidelberg New York
ISBN-13   978-3-540-70955-8 Springer Berlin Heidelberg New York

Springer is a part of Springer Science+Business Media

springer.com

© Springer-Verlag Berlin Heidelberg 2008
Printed in Germany

Typesetting: Camera-ready by author, data conversion by Scientific Publishing Services, Chennai, India
Printed on acid-free paper   SPIN: 12438815   06/3180   5 4 3 2 1 0

Lecture Notes in Computer Science 4950

Commenced Publication in 1973
Founding and Former Series Editors:
Gerhard Goos, Juris Hartmanis, and Jan van Leeuwen

Editorial Board

Preface

From May 28 to June 1, 2007, a seminar on "Information Visualization – Human-Centered Issues in Visual Representation, Interaction, and Evaluation" (♯07221) took place at the International Conference and Research Center for Computer Science, Dagstuhl Castle, Germany. The center was initiated by the German government to promote informatics research at an international level. It seeks to foster dialog among the computer science research community, advance academic education and professional development, and transfer knowledge between academia and industry.

The primary feature of Dagstuhl is its series of week-long seminars on various topics in computer science. Dagstuhl seminars are frequently described as being the most productive academic events that the participant researchers have ever experienced. The informal and friendly atmosphere fostered at the center promotes personal interaction between the guests. Traditionally, there is no set program followed at Dagstuhl seminars. Instead, the pace and the procedure are determined by the presentations offered during the seminar and the discussion results. Further general information about Dagstuhl seminars can be found on the Dagstuhl Castle webpage[1].

Information visualization (InfoVis) is a relatively new research area, which focuses on the use of visualization techniques to help people understand and analyze data. While related fields such as scientific visualization involve the presentation of data that has some physical or geometric correspondence, information visualization centers on abstract information without such correspondences, i.e., information that cannot be mapped into the physical world in most cases. Examples of such abstract data are symbolic, tabular, networked, hierarchical, or textual information sources. The ever increasing amount of data generated or made available every day amplifies the urgent need for InfoVis tools. To give visualization a firm fundament, InfoVis combines several aspects of different research areas, such as scientific visualization, human-computer interaction, data mining, information design, cognitive psychology, visual perception, cartography, graph drawing, and computer graphics.

One main goal of our seminar on information visualization was to bring together researchers and practitioners from the above-mentioned research areas as well as from application areas, such as bioinformatics, finance, the geosciences, software engineering, and telecommunications. Several international conferences include information visualization topics, each with a slightly different high-level objective. Another goal of the Dagstuhl seminar was to consolidate these diverse areas in one joint meeting. The seminar allowed critical reflection on actual research efforts, the state of the field, evaluation challenges, and future directions. A detailed report on participation, program, and discussions was published in the journal *Information Visualization*, published by Palgrave Macmillan Ltd. (6(3):189-196, 2007).

---

[1] http://www.dagstuhl.de/en/about-dagstuhl/

This book is the outcome of Dagstuhl Seminar ♯07221. It documents and extends the findings and discussions of the various sessions in detail. During the last day of the seminar, the most important topics for publication were identified and assigned to those interested. The resulting author groups worked together to write papers on the chosen topics. In the following, we briefly present the different papers.

Book Structure

Paper 1 "The Value of Information Visualization", provides a discussion of issues surrounding the utility and benefits of InfoVis. The paper identifies why communicating the value of InfoVis is more difficult than in many other areas, and it provides a number of arguments and examples that help to illustrate InfoVis' value.

Paper 2 "Evaluating Information Visualizations", discusses the challenges associated with the evaluation of InfoVis tools and approaches. Different types of evaluation are described as well as the advantages and disadvantages of different empirical methodologies.

Paper 3 "Theoretical Foundations of Information Visualization", addresses an important issue: InfoVis, being related to many other diverse disciplines, suffers from not being based on a clear underlying theory. Drawing on theories within associated disciplines, three different approaches to theoretical foundations of information visualization are presented: data-centric predictive theory, information theory, and scientific modeling.

Paper 4 "Teaching Information Visualization", presents the results of a survey about InfoVis-related courses that was distributed to the Dagstuhl attendees during the seminar. It also summarizes the discussions about teaching held by the attendees during a special session on that topic. The paper includes the perspectives of three seminar participants in relation to their own InfoVis-related teaching experiences.

Paper 5 "Creation and Collaboration: Engaging New Audiences for Information Visualization", discusses creation and collaboration tools for interactive visualization. The paper characterizes the increasingly diverse audience for visualization technology, and it formulates a design space for new creative and collaborative tools to support these users.

Paper 6 "Process and Pitfalls in Writing Information Visualization Research Papers", identifies a set of pitfalls and problems that recur in many InfoVis-related papers, using a chronological model of the research process. The aim of this paper is to help authors avoid these pitfalls and write better papers. Reviewers might also find these pitfalls interesting to consider when evaluating the merits of a paper.

Paper 7 "Visual Analytics: Definition, Process, and Challenges", describes the related and growing field of visual analytics. The paper explains the perceived difference between visual analytics and InfoVis, and it identifies the technical challenges faced by visual analytics researchers. The paper concludes by describing a number of visual analytics applications.

Acknowledgments

We would like to thank all those who participated in the seminar for the lively discussions as well as the scientific directorate of Dagstuhl Castle for giving us the opportunity to organize this event. The abstracts and talks can be found on the Dagstuhl website for this seminar[2]. In addition, we are also grateful to all the authors for their valuable time and contributions to the book. Last but not least, the seminar would not have been possible without the great help of the staff of Dagstuhl Castle. We would like to thank all of them for their assistance.

March 2008

Andreas Kerren
John T. Stasko
Jean-Daniel Fekete
Chris North

---

[2] http://www.dagstuhl.de/07221

Table of Contents

Part I. General Reflections

Part II. Specific Aspects

The Value of Information Visualization

Jean-Daniel Fekete[1], Jarke J. van Wijk[2], John T. Stasko[3], and Chris North[4]

[1] Université Paris-Sud, INRIA, Bât 490,
F-91405 Orsay Cedex, France,
Jean-Daniel.Fekete@inria.fr,
http://www.aviz.fr/~fekete/
[2] Department of Mathematics and Computing Science,
Eindhoven University of Technology, P.O. Box 513,
5600 MB EINDHOVEN, The Netherlands,
vanwijk@win.tue.nl,
http://www.win.tue.nl/~vanwijk/
[3] School of Interactive Computing, College of Computing & GVU Center,
Georgia Institute of Technology, 85 5th St., NW,
Atlanta, GA 30332-0760, USA,
stasko@cc.gatech.edu,
http://www.cc.gatech.edu/~john.stasko
[4] Dept of Computer Science, 2202 Kraft Drive,
Virginia Tech, Blacksburg, VA 24061-0106, USA,
north@vt.edu,
http://people.cs.vt.edu/~north/

Abstract. Researchers and users of Information Visualization are convinced that it has value. This value can easily be communicated to others in a face-to-face setting, such that this value is experienced in practice. To convince broader audiences, and also, to understand the intrinsic qualities of visualization is more difficult, however. In this paper we consider information visualization from different points of view, and gather arguments to explain the value of our field.

1 Problems and Challenges

This paper provides a discussion of issues surrounding the value of Information Visualization (InfoVis). The very existence of the paper should alert the reader that challenges do exist in both recognizing and communicating the field's value. After all, if the value would be clear and undisputed, there would be no need to write the paper! Unfortunately, the current situation is far from that. By its very focus and purpose, InfoVis is a discipline that makes the recognition of value extremely difficult, a point that will be expanded below.

Why is showing value important? Well, today's research environment places great importance on evaluation involving quantifiable metrics that can be assessed and judged with clarity and accuracy. Organizations sponsoring research and corporations that serve to benefit from it want to know that the monetary investments they make are being well-spent. Researchers are being challenged

A. Kerren et al. (Eds.): Information Visualization, LNCS 4950, pp. 1–18, 2008.

to show that their inventions are measurably better than the existing state of the art.

In broad analytic fields, of which we include InfoVis as a member, the existence of a ground truth for a problem can greatly facilitate evaluations of value. For instance, consider the field of computer vision and algorithms for identifying objects from scenes. It is very easy to create a library of images upon which new algorithms can be tested. From that, one can measure how well each algorithm performs and compare results precisely. The TREC [29] and MUC [3] Contests are examples of this type of evaluation.

Even with a human in the loop, certain fields lend themselves very well to quantifiable evaluations. Consider systems that support search for particular documents or facts. Even though different people will perform differently using a system, researchers can run repeated search trials and measure how often a person is able to find the target and how long the search took. Averaged over a large number of human participants, this task yields quantifiable results that can be measured and communicated quite easily. People or organizations then using the technology can make well-informed judgments about the value of new tools.

So why is identifying the value of InfoVis so difficult? To help answer that question, let us turn to what is probably the most accepted definition of InfoVis, one that comes from Card, Mackinlay, and Shneiderman and that actually is their definition for "visualization." They describe visualization as "the use of computer-supported, interactive visual representations of data to amplify cognition." [2] The last three words of their definition communicate the ultimate purpose of visualization, to amplify cognition. So, returning to our discussion above, is the amplification of cognition something with a ground truth that is easily and precisely measurable? Clearly it is not and so results the key challenge in communicating the value of InfoVis.

Further examining the use and purpose of InfoVis helps understand why communicating its value is so difficult. InfoVis systems are best applied for exploratory tasks, ones that involve browsing a large information space. Frequently, the person using the InfoVis system may not have a specific goal or question in mind. Instead, the person simply may be examining the data to learn more about it, to make new discoveries, or to gain insight about it. The exploratory process itself may influence the questions and tasks that arise.

Conversely, one might argue that when a person does have a specific question to be answered, InfoVis systems are often not the best tools to use. Instead, the person may formulate his or her question into a query that can be dispatched to a database or to a search engine that is likely to provide the answer to that precise question quickly and accurately.

InfoVis systems, on the other hand, appear to be most useful when a person simply does not know what questions to ask about the data or when the person wants to ask better, more meaningful questions. InfoVis systems help people to rapidly narrow in from a large space and find parts of the data to study more carefully.

Unfortunately, however, activities like exploration, browsing, gaining insight, and asking better questions are not ones that are easily amenable to establishing and measuring a ground truth. This realization is at the core of all the issues involved in communicating the value of InfoVis. By its very nature, by its very purpose, InfoVis presents fundamental challenges for identifying and measuring value. For instance, how does one measure insight? How does one quantify the benefits of an InfoVis system used for exploring an information space to gain a broad understanding of it? For these reasons and others, InfoVis is fundamentally challenging to evaluate [17].

If we accept that InfoVis may be most valuable as an exploratory aid, then identifying situations where browsing is useful can help to determine scenarios most likely to illustrate InfoVis' value. Lin [11] describes a number of conditions in which browsing is useful:

- When there is a good underlying structure so that items close to one another can be inferred to be similar
- When users are unfamiliar with a collection's contents
- When users have limited understanding of how a system is organized and prefer a less cognitively loaded method of exploration
- When users have difficulty verbalizing the underlying information need
- When information is easier to recognize than describe

These conditions serve as good criteria for determining situations in which the value of InfoVis may be most evident.

1.1 Epistemological Issues

Natural sciences are about understanding how nature works. Mathematics is about truth and systems of verifiable inferences. Human sciences are about understanding Man in various perspectives. Information Visualization is about developing insights from collected data, not about understanding a specific domain. Its object is unique and therefore raises interest and skepticism.

Science has focused on producing results: the goal was essentially the creation and validation of new theories compatible with collected facts. The importance of the process — coined as the Method — was raised by the development of *epistemology* in the 20th century, in particular with the work of Karl R. Popper (1902–1994) [18]. It showed that the Method was paramount to the activity of science.

Karl Popper has explained that a scientific theory cannot be proved true, it can only be *falsified*. Therefore, a scientific domain searches for theories that are as compatible as possible with empirical facts. The good theories are the ones that have been selected by domain experts among a set of competing theories in regard of the facts that they should describe. Popper considers science as a Darwinian selection process among competing theories.

Still, no other scientific domain has argued that generating insights was important for science. Popper does not explain how a new theory emerges; he only explains how it is selected when it emerges. Furthermore, Popper has demonstrated in an article called "The Problem of Induction" that new theories cannot rationally emerge from empirical data: it is impossible to justify a law by observation or experiment, since it 'transcends experience'.

Information Visualization is still an inductive method in the sense that it is meant at generating new insights and ideas that are the seeds of theories, but it does it by using human perception as a very fast filter: if vision perceives some pattern, there might be a pattern in the data that reveals a structure. Drilling down allows the same perception system to confirm or infirm the pattern very quickly. Therefore, information visualization is meant at "speeding up" the process of filtering among competing theories regarding collected data by relying on the speed of the perception system. Therefore, it plays a special role in the sciences as an insight generating method. It is not only compatible with Popper's epistemology system but it furthermore provides a mechanism for accelerating its otherwise painful Darwinian selection process.

1.2 Moving Forward

It is clear that InfoVis researchers and practitioners face an important challenge in communicating the value of InfoVis. In the remainder of the paper we explore this challenge more deeply and we provide several answers to the questions "How and why is InfoVis useful?". Since there are several audiences to convince, we present a number of different sections that are each facets of argumentation to explain why InfoVis is useful and effective as a mean of understanding complex datasets and developing insights. The contents of the sections are gathered from practitioners who already attested that the arguments developed were convincing. We hope they will be useful to you as well.

2 Cognitive and Perceptual Arguments

Several famous historical figures have argued that the eye was the main sense to help us understand nature.

> The eye. . .
> the window of the soul,
> is the principal means
> by which the central sense
> can most completely and
> abundantly appreciate
> the infinite works of nature.
>
> Leonardo da Vinci (1452 – 1519)

Leonardo's words are inspirational and they are echoed in everyday expressions that we are all familiar with such as, "Seeing is believing" and "A picture is worth a thousand words." Is there actual support for these sentiments, however?

Let us first consider the case of the phrase, "A picture is worth a thousand words." While people may agree or disagree with the sentiments behind that cliché, specific examples can help support the claim. Consider, for instance, the

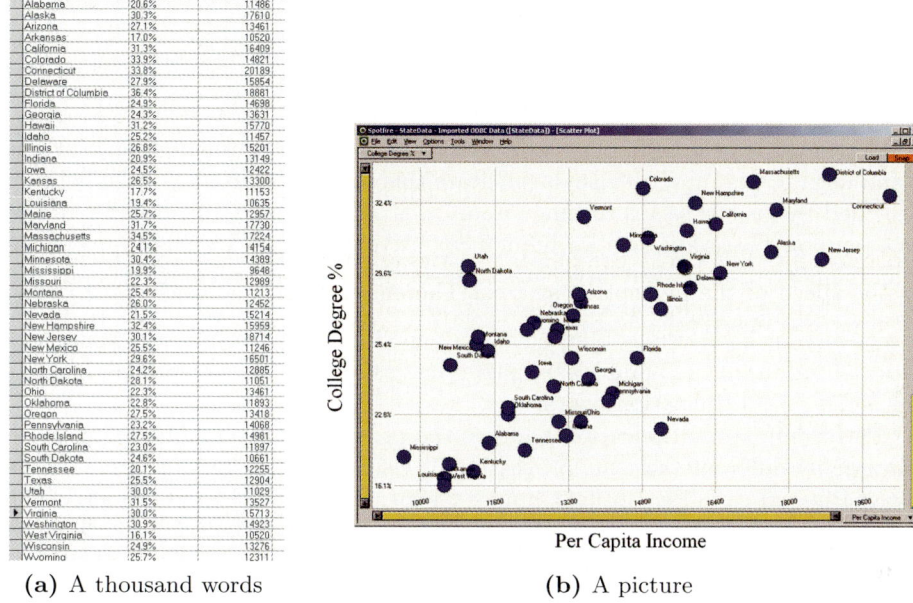

State	College Degree %	Per Capita Income
Alabama	20.6%	11486
Alaska	30.3%	17610
Arizona	27.1%	13461
Arkansas	17.0%	10520
California	31.3%	16409
Colorado	33.9%	14821
Connecticut	33.8%	20189
Delaware	27.9%	15854
District of Columbia	36.4%	18881
Florida	24.3%	14698
Georgia	24.3%	13631
Hawaii	31.2%	15770
Idaho	25.2%	11457
Illinois	26.8%	15201
Indiana	20.9%	13149
Iowa	24.5%	12422
Kansas	26.5%	13300
Kentucky	17.7%	11153
Louisiana	19.4%	10635
Maine	25.7%	12957
Maryland	31.7%	17730
Massachusetts	34.5%	17224
Michigan	24.1%	14154
Minnesota	30.4%	14389
Mississippi	19.9%	9648
Missouri	22.3%	12989
Montana	25.4%	11213
Nebraska	26.0%	12452
Nevada	21.5%	15214
New Hampshire	32.4%	15959
New Jersey	30.1%	18714
New Mexico	25.5%	11246
New York	29.6%	16501
North Carolina	24.2%	12885
North Dakota	28.1%	11051
Ohio	22.3%	13461
Oklahoma	22.8%	11893
Oregon	27.5%	13418
Pennsylvania	23.2%	14068
Rhode Island	27.5%	14981
South Carolina	23.0%	11897
South Dakota	24.6%	10661
Tennessee	20.1%	12255
Texas	25.5%	12904
Utah	30.0%	11029
▶ Vermont	31.5%	13527
Virginia	30.0%	15713
Washington	30.9%	14923
West Virginia	16.1%	10520
Wisconsin	24.9%	13276
Wyoming	25.7%	12311

(a) A thousand words   **(b)** A picture

Fig. 1. "A picture is worth a thousand words"

example shown in Figure 1. Part (a) shows a spreadsheet with data for the 50 states and the District of Columbia in the U.S. Also shown are the percentage of citizens of each state with a college degree and the per capita income of the states' citizens.

Given just the spreadsheet, answering a question such as, "Which state has the highest average income?" is not too difficult. A simple scan of the income column likely will produce the correct answer in a few seconds. More complex questions can be quite challenging given just the data, however. For example, are the college degree percentage and income correlated? If they are correlated, are there particular states that are outliers to the correlation? These questions are much more difficult to answer using only the spreadsheet.

Now, let us turn to a graphical visualization of the data. If we simply draw the data in a scatterplot as shown in part (b), the questions now become much easier to answer. Specifically, there does appear to be an overall correlation between the two attributes and states such as Nevada and Utah are outliers on the correlation. The simple act of plotting the spreadsheet data in this more meaningfully communicative form makes these kinds of analytic queries easier to answer correctly and more rapidly.

Note that the spreadsheet itself is a visual representation of the data that facilitates queries as well. Consider how difficult the three questions would be if the data for each state was recorded on a separate piece of paper or webpage. Or worse yet, what if the data values were read to you and you had to answer the questions? In this case, already challenging questions become practically impossible.

2.1   Cognitive Benefits

While the states example illustrates that visualizations can help people understand data better, how do visuals facilitate this process? The core of the benefits provided by visuals seems to hinge upon their acting as a frame of reference or as a temporary storage area for human cognitive processes. Visuals augment human memory to provide a larger working set for thinking and analysis and thus become external cognition aids. Consider the process of multiplying two long integers in your head versus then having a pencil and paper available. The visual representations of the numbers on paper acts as a memory aid while performing the series of multiplication operations.

Performing a multiplication requires the processing of symbolic data, which is arguably different than the processing of visual features and shapes. In "Abstract Planning and Perceptual Chunks: Elements of Expertise in Geometry" [8], Koedinger and Anderson show that experts in geometry effectively use their vision to infer geometrical properties (parallelism, connectivity, relative positions) on diagrams; they solve simple problems quickly and accurately, several of magnitude faster than if they had to apply symbolic inference rules.

Larkin and Simon, in their landmark paper "Why a diagram is (sometimes) worth 10,000 words" [10], discuss how graphical visualization can support more efficient task performance by allowing substitution of rapid perceptual inferences for difficult logical inferences and by reducing the search for information required for task completion. They do note that text can be better than graphics for certain tasks, however.

Don Norman provides many illustrative examples where visuals can greatly assist task performance and efficiency [15]. He argues that it is vital to match the representation used in a visualization to the task it is addressing. The examples he cites show how judicious visuals can aid information access and computation.

Earlier, we noted how the definition of visualization from Card, Mackinlay and Shneiderman [2] focused on the use of visuals to "amplify cognition." Following that definition, the authors listed a number of key ways that the visuals can amplify cognition:

- Increasing memory and processing resources available
- Reducing search for information
- Enhancing the recognition of patterns
- Enabling perceptual inference operations
- Using perceptual attention mechanisms for monitoring
- Encoding info in a manipulable medium

2.2   Perceptual Support

Most lectures on Information Visualization argue about theoretical properties of the visual system or more broadly to the perception abilities of humans. Rational arguments rely on information theory [21] and psychological findings.

According to *Information Theory*, vision is the sense that has the largest bandwidth: 100 Mb/s [30]. Audition only has around 100 b/s. In that respect, the visual canal is the best suited to carrying information to the brain.

According to Ware [30], there are two main *psychological theories* that explain how vision can be used effectively to perceive features and shapes. At the low level, *Preattentive processing* theory [23] explains what visual features can be effectively processed. At a higher cognitive level, the *Gestalt* theory [9] describes some principles used by our brain to understand an image.

Preattentive processing theory explains that some visual features can be perceived very rapidly and accurately by our low-level visual system. For example, when looking at the group of blue circles in Figure 2, it takes no time and no effort to see the red circle in the middle. It would be as easy and fast to see that there is no red circle, or to evaluate the relative quantity of red and blue circles. Color is one type of feature that can be processed preattentively, but only for some tasks and within some limits. For example, if there were more than seven colors used in Figure 2, answering the question could not be done with preattentive processing and would require sequential scanning, a much longer process.

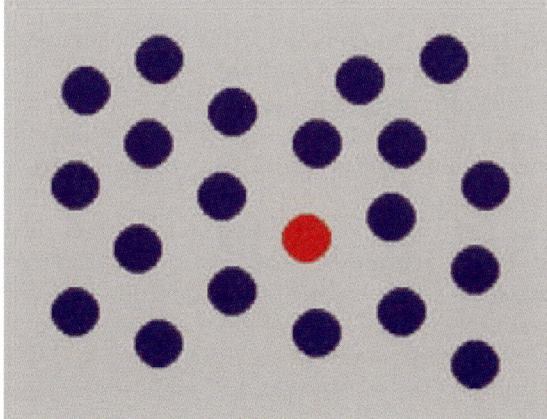

Fig. 2. Example of preattentively processed task: finding if there is a red circle among the blue circles

There is a long list of visual features that can be preattentively processed for some tasks, including line orientation, line length or width, closure, curvature, color and many more. Information visualization relies on this theory to choose the visual encoding used to display data to allow the most interesting visual queries to be done preattentively.

Gestalt theory explains important principles followed by the visual system when it tries to understand an image. According to Ware [30], it is based on the following principles:

Proximity Things that are close together are perceptually grouped together;

Similarity Similar elements tend to be grouped together;

Continuity Visual elements that are smoothly connected or continuous tend to be grouped;

Symmetry Two symmetrically arranged visual elements are more likely to be perceived as a whole;

Closure A closed contour tends to be seen as an object;

Relative Size Smaller components of a pattern tend to be perceived as objects whereas large ones as a background.

Information Visualization experts design visual representations that try to follow these principles. For example, graph layout algorithms such as [14] designed to find communities in social networks adhere to the *proximity* principle by placing nodes that are connected to a dense group close together and push away nodes that are connected to another dense group. The Treemap algorithm [5] uses the *closure* principle to layout a tree: children of a node are placed inside their parent node.

3   Success Stories

Information Visualization is much easier to explain using demonstrations than words. However, to be understood, the data used should be familiar to the audience and interesting. Preparing demonstrations targeted at all the possible audiences is not possible but there are some datasets that interest most audiences and greatly help make the point. Several striking static examples can be found in Tufte's books [24,25,26].

To better explain the value of visualization, demonstrations should start using a simple question, show that a good representation answers the question at once and then argue about additional benefits, *i.e.* questions the users did not knew they had. From the users perspective, a good representation will confirm what they already know, let them answer at once the question asked and show them several insights, leading to the so-called "a-ha" moments when they feel like they understand the dataset.

3.1   Striking Examples

Static examples used by most InfoVis courses include the map of Napoleon's 1812 March on Moscow drawn in 1869 by M. Minard (Figure 3) and the map of London in 1854 overlaid with marks positioning cholera victims that led John Snow to discovering the origin of the epidemic: infected water extracted with a water pump at the center of the marks (Figure 4).

In general, good examples show known facts (although sometimes forgotten) and reveal several unexpected insights at once. Minard's map can help answer the question: "What were the casualties of Napoleon's Russian invasion in 1812?". The map reveals at once the magnitude of casualties (from 400,000 to 10,000

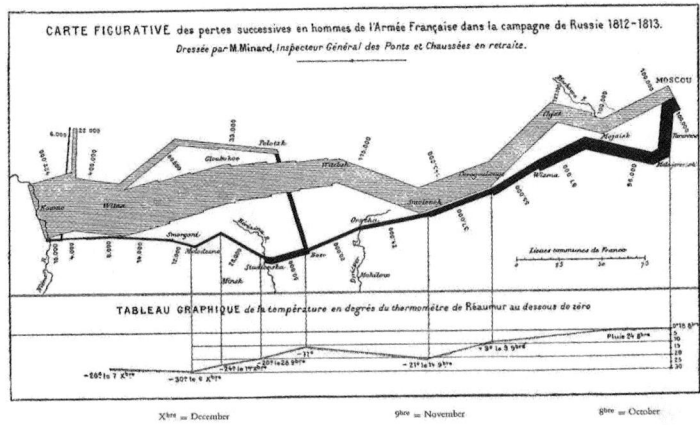

Fig. 3. Napoleon's March on Moscow depicted by M. Minard [12]. Width indicates the number of soldiers. Temperature during the retreat is presented below the map. *Image courtesy of École Nationale des Ponts et Chaussées.*

soldiers) as well as the devastating effect of crossing the Berezina river (50,000 soldiers before, 25,000 after). The depiction confirms that Napoleon lost the invasion (a well known fact) and reveals many other facts, such as the continuous death rate due to disease and the "scorched earth" tactics of Russia instead of specific death tolls of large battles.

John Snow's map was made to answer the question: "What is the origin of the London cholera epidemics?". Contrary to the previous map, the answer requires some thinking. Black rectangles indicate location of deaths. At the center of the infected zone lies a water pump that John Snow found to be responsible for the infection. Once again, choosing the right representation was essential for finding the answer. As a side-effect, the map reveals the magnitude of the epidemic.

Figure 1 answers the question: "Is there a relationship between income and college degree?" by showing a scatter plot of income by degree for each US state. The answer is the obvious: yes, but there is much more. There seems to be a linear correlation between them and some outliers such as Nevada (likely due to Las Vegas) and Utah do exist, raising new unexpected questions.

Information Visualization couples interaction and visual representation so its power is better demonstrated interactively. The simplest demonstration suited to the largest audience is probably the Dynamic HomeFinder[5] [32] . It shows the map of the Washington D.C. area overlaid with all the homes for sale (Figure 5). Dynamic queries implemented by sliders and check-boxes interactively filter-out homes that do not fit specific criteria such as cost or number of bedrooms.

Using the interactive controls, it becomes easy to find homes with the desired attributes or understand how attributes' constraints should be relaxed to find some matching homes. Unexpectedly, the Dynamic HomeFinder also reveals the

---

[5] http://www.cs.umd.edu/hcil/pubs/dq-home.zip

Fig. 4. Illustration of John Snow's deduction that a cholera epidemic was caused by a bad water pump, circa 1854 [4]. Black rectangles indicate location of deaths.

unpopular neighborhoods around Washington D.C. since they are places where the homes are cheaper, and the wealthy ones where the houses are more expensive.

Many more examples can be found to demonstrate that InfoVis is effective. The Map of the Market[6], represented by a squarified treemap, is interesting for people holding stocks or interested by economic matters. InfoZoom video on the analysis of Formula 1 results[7] is interesting for car racing amateurs. The video[8] comparing two large biological classification trees is interesting to some biologists. The Baby Name Wizard's NameVoyager[9] is useful for persons searching a name for their baby to come and a large number of other persons as witnessed by [31].

With the advent of Social InfoVis through web sites such as Swivel[10] or IBM's Many-Eyes[11], more examples can be found to convince specific audiences. Still, the process of explaining how InfoVis works remains the same: ask a question that interests people, show the right representation, let the audience understand the representation, answer the question and realize how many more unexpected findings and questions arise.

---

[6] http://www.smartmoney.com/marketmap/

[7] http://www.infozoom.com/enu/infozoom/video.htm

[8] http://www.fit.fraunhofer.de/~cici/InfoVis2003/StandardForm/Flash/
InfoZoomTrees.html

[9] http://babynamewizard.com/namevoyager/

[10] http://www.swivel.com

[11] http://www.many-eyes.com

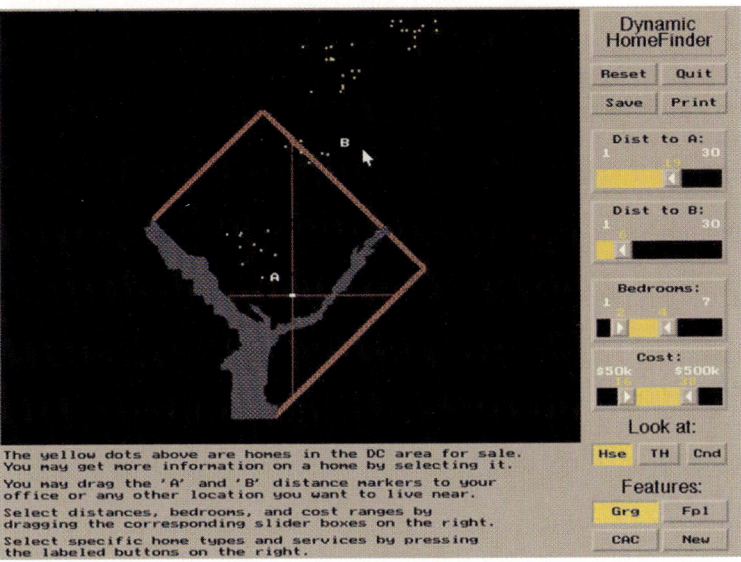

Fig. 5. Dynamic HomeFinder showing the Washington D.C. area with homes available for sale and controls to filter them according to several criterion.

3.2 Testimonials

One effective line of argumentation about the value of InfoVis is through reporting the success of projects that used InfoVis techniques. These stories exist but have not been advertised in general scientific publications until recently [20,16,13]. One problem with trying to report on the success of a project is that visualization is rarely the only method used to reach the success. For example, in biological research, the insights gained by an InfoVis system can lead to an important discovery that is difficult to attribute mainly to the visualization since it also required months of experimentation to verify the theory formulated from the insights. In fact, most good human-computer interaction systems allow users to forget about the system and focus on their task only, which is probably one reason why success stories are not so common in the InfoVis literature.

Besides these stories that are empirical evidence of the utility of information visualization, there are strong theoretical arguments to how and why information visualization works.

4 Information Visualization vs. Automatic Analysis

Several scientific domains are concerned by understanding complex data. Statistics is the oldest, but Data Mining — a subfield of Artificial Intelligence — is also concerned with automatically understanding the structure of data. Therefore, InfoVis practitioners frequently need to explain what InfoVis can do that statistics and data mining cannot.

4.1   Statistics

Statistics is a well grounded field but is composed of several subfields such as descriptive statistics, classical statistics (also called *confirmatory* statistics), Bayesian statistics and Exploratory Data Analysis. Information Visualization is sometimes considered as a descendant and expansion of Exploratory Data Analysis.

The differences between these subfields are the methods and the nature of the answer they seek. All of them start with a problem and gathered data that is related to the problem to solve. Classical analysis starts by designing a *model* of the data, then uses mathematical analysis to test whether the model is refuted or not by the data to conclude positively or negatively. The main challenge for classical statistics is to find a model.

Exploratory Data Analysis performs an analysis using visual methods to acquire insights of what the data looks like, usually to find a model. It uses visual exploration methods to get the insights.

So why is visualization useful before the modeling? Because, there are cases when we have no clear idea on the nature of the data and have no model.

To show why visualization can help finding a model, Anscombe in [1] has designed four datasets that exhibit the same statistical profile but are quite different in shape, as shown in Figure 6. They have the following characteristics[12]:

- mean of the x values = 9.0
- mean of the y values = 7.5
- equation of the least-squared regression line is: $y = 3 + 0.5x$
- sums of squared errors (about the mean) = 110.0
- regression sums of squared errors (variance accounted for by x) = 27.5
- residual sums of squared errors (about the regression line) = 13.75
- correlation coefficient = 0.82
- coefficient of determination = 0.67.

Visualization is much more effective at showing the differences between these datasets than statistics. Although the datasets are synthetic, Anscombe's Quartet demonstrates that looking at the shape of the data is sometimes better than relying on statistical characterizations alone.

4.2   Data Mining

More than statistics, the goal of data mining is to automatically find interesting facts in large datasets. It is thus legitimate to wonder whether data mining, as a competitor of InfoVis, can overcome and replace the visual capacity of humans.

This question has been addressed by Spence and Garrison in [22] where they describe a simple plot called the Hertzsprung Russell Diagram (Figure 7a). It represents the temperature of stars on the X axis and their magnitude on the Y axis. Asking a person to summarize the diagram produces Figure 7b. It

---

[12] See `http://astro.swarthmore.edu/astro121/anscombe.html` for details

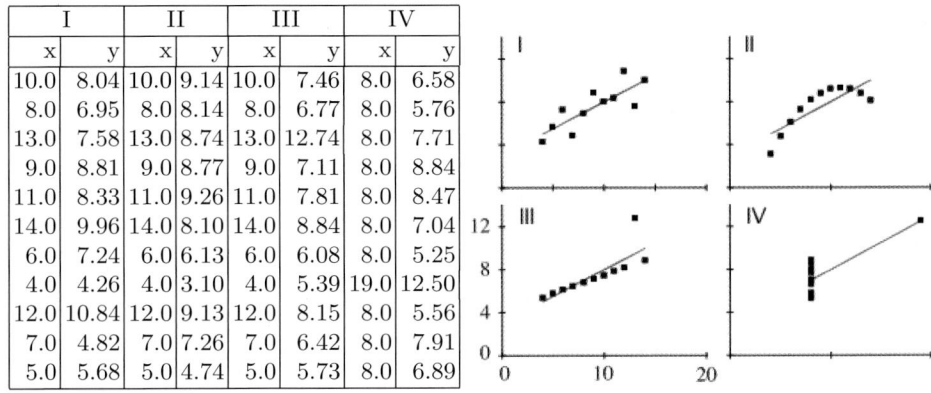

I		II		III		IV	
x	y	x	y	x	y	x	y
10.0	8.04	10.0	9.14	10.0	7.46	8.0	6.58
8.0	6.95	8.0	8.14	8.0	6.77	8.0	5.76
13.0	7.58	13.0	8.74	13.0	12.74	8.0	7.71
9.0	8.81	9.0	8.77	9.0	7.11	8.0	8.84
11.0	8.33	11.0	9.26	11.0	7.81	8.0	8.47
14.0	9.96	14.0	8.10	14.0	8.84	8.0	7.04
6.0	7.24	6.0	6.13	6.0	6.08	8.0	5.25
4.0	4.26	4.0	3.10	4.0	5.39	19.0	12.50
12.0	10.84	12.0	9.13	12.0	8.15	8.0	5.56
7.0	4.82	7.0	7.26	7.0	6.42	8.0	7.91
5.0	5.68	5.0	4.74	5.0	5.73	8.0	6.89

(a) Four datasets with different values and the same statistical profile

(b) Dot Plot of the four datasets

Fig. 6. Anscombe's Quartet

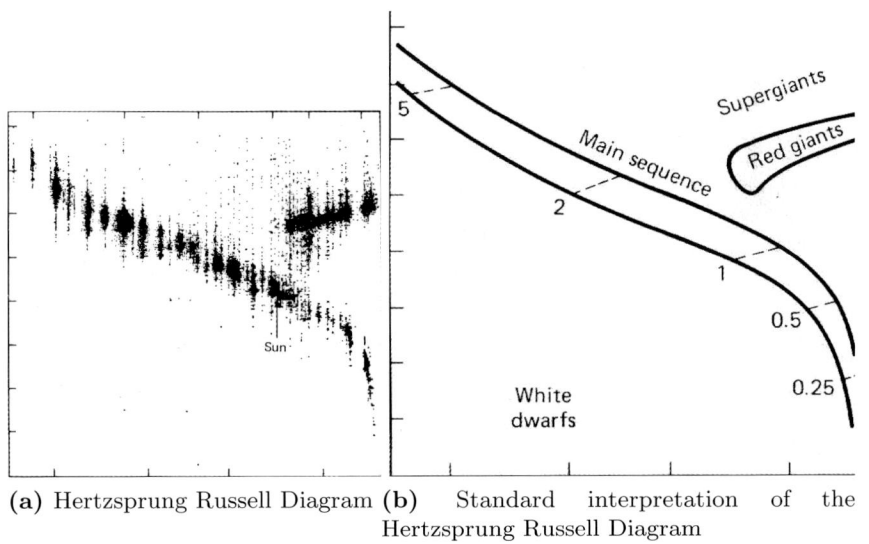

(a) Hertzsprung Russell Diagram

(b) Standard interpretation of the Hertzsprung Russell Diagram

Fig. 7. Hertzsprung Russell Diagram and its standard interpretation

turns out that no automatic analysis method has been able to find the same summarization, due to the noise and artifacts on the data such as the vertical bands.

Our vision system has evolved with the human specie to help us survive in a hostile world. We train it to avoid obstacles since we learn how to walk. It remains remarkably effective at filtering-out noise from useful data, a very important capability for hunters in deep forests to distinguish the prey moving behind leaves. We have relied on it and trained it to survive for millennia and it still surpasses automatic data mining methods to spot interesting patterns. Data mining still needs to improve to match these pattern matching capabilities.

4.3   Automating or Not?

Is there a competition between confirmatory, automated and exploratory methods? No, they answer different questions. When a model is known in advance or expected, using statistics is the right method. When a dataset becomes too large to be visualized directly, automating some analysis is required. When exploring a dataset in search of insights, information visualization should be used, possibly in conjunction with data mining techniques if the dataset is too large.

Furthermore, combining data mining with visualization is the central issue of *Visual Analytics*, described by the paper *Visual Analytics: Definition, Process, and Challenges* in this book [6].

5   An Economical Model of Value

One important question is how to assess the value of visualization, ranging from the evaluation of one specific use-case to the discipline in general. If we know how to do this, then this might lead to an assessment of the current status as well as the identification of success factors. An attempt was given by Van Wijk [27] and is summarized here. After a short overview of his model, we discuss how this model can be applied for InfoVis.

Visualization can be considered as a technology, a collection of methods, techniques, and tools developed and applied to satisfy a need. Hence, standard technological measures apply: Visualization has to be effective and efficient. To measure these, an economic point of view is adopted. Instead of trying to understand why visualization works (see previous sections), here visualization is considered from the outside, and an attempt is made to measure its profit. The profit of visualization is defined as the difference between the value of the increase in knowledge and the costs made to obtain this insight. Obviously, in practice these are hard to quantify, but it is illuminating to attempt so. A schematic model is considered: One visualization method V is used by n users to visualize a data set m times each, where each session takes k explorative steps. The value of an increase in knowledge (or insight) has to be judged by the user. Users can be satisfied intrinsically by new knowledge, as an enrichment of their understanding of the world. A more pragmatic and operational point of view is to consider if the new knowledge influences decisions, leads to actions, and, hopefully, improves the quality of these. The overall gain now is $nm(W(\Delta K))$, where $W(\Delta K))$ represents the value of the increase in knowledge.

Concerning the costs for the use of (a specific) visualization V, these can be split into various factors. Initial research and development costs C_i have to be made; a user has to make initial costs C_u, because he has to spend time to select and acquire V, and understand how to use it; per session initial costs C_s

have to be made, such as conversion of the data; and finally during a session a user makes costs C_e, because he has to spend time to watch and understand the visualization, and interactively explore the data set. The overall profit now is

$$F = nm(W(\Delta K) - C_s - kC_e) - C_i - nC_u.$$

In other words, this leads to the obvious insight that a great visualization method is used by many people, who use it routinely to obtain highly valuable knowledge, while having to spend little time and money on hardware, software, and effort. And also, no alternatives that are more cost-effective should be available.

In the original paper a number of examples of more or less successful visualization methods are given, viewed in terms of this model. One InfoVis application was considered: SequoiaView, a tool to visualize the contents of a hard disk, using cushion treemaps [28]. The popularity of this tool can be explained from the concrete and useful insights obtained, as well as the low costs in all respects associated with its application.

When we consider InfoVis in general, we can also come to positive conclusions for almost all parameters, and hence predict a bright future for our field. The number of potential users is very large. Data in the form of tables, hierarchies, and networks is ubiquitous, as well as the need to get insight in these. This holds for professional applications, but also for private use at home. Many people have a need to get an overview of their email, financial transfers, media collections, and to search in external data bases, for instance to find a house, vacation destination, or another product that meets their needs. Methods and techniques from InfoVis, in the form of separate tools or integrated in custom applications, can be highly effective here to provide such overviews. Also, many of these activities will be repeated regularly, hence both n and m are high. The growing field of Casual InfoVis [19] further illustrates how InfoVis techniques are becoming more common in people's everyday lives.

The costs C_e that have to be made to understand visualizations depend on the prior experience of the users as well as the complexity of the imagery shown. On the positive side, the use of graphics to show data is highly familiar, and bar-charts, pie-charts, and other forms of business graphics are ubiquitous. On the other hand, one should not overestimate familiarity. The scatterplot seems to be at the boundary: Considered as trivial in the InfoVis community, but too hard to understand (if the horizontal axis does not represent time) by a lay-audience, according to Matthew Ericson, deputy graphics director of the New York Times in his keynote presentation at IEEE InfoVis 2007. Visual literacy is an area where more work can be done, but on the other hand, InfoVis does have a strong edge compared to non-visual methods here. And, there are examples of areas where complex visual coding has been a great success, with the invention of the script as prime example.

The costs C_s per session and C_u per user can be reduced by tight integration with applications. The average user will not be interested in producing visualizations, her focus will be on solving her own problem, where visualization is one of the means to this end. Separate InfoVis tools are useful for specialists, which use them on a day-to-day basis. For many other users, integration within their favourite tool is much more effective. An example of an environment that offers such a tight integration is the ubiquitous spreadsheet, where storage, manipulation, and presentation of data are offered; or the graphs and maps shown on many web sites (and newspapers!) to show data. From an InfoVis point of view, the presentations offered here can often be improved, and also, the interaction provided is often limited. Nevertheless, all these examples acknowledge the value of visualization for many applications.

The initial costs C_i for new InfoVis methods and techniques roughly fall into two categories: Research and Development. Research costs can be high, because it is often hard to improve on the state of the art, and because many experiments (ranging from the development of prototypes to user experiments) are needed. On the other hand, when problems are addressed with many potential usages, these costs are still quite limited. Development costs can also be high. It takes time and effort to produce software that is stable and useful under all conditions, and that is tightly integrated with its context, but here also one has to take advantage of the large potential market. Development and availability of suitable middleware, for instance as libraries or plug-ins that can easily customized for the problem at hand is an obvious route here.

One intriguing aspect here is how much customization is needed to solve the problem concerned. On one hand, in many applications one of the standard data types of InfoVis is central (table, tree, graph, text), and when the number of items is not too high, the problem is not too hard to solve. On the other hand, for large numbers of items one typically has to exploit all a priori knowledge of the data set and tune the visualization accordingly; also, for applications such as software visualizations all these data types pop up simultaneously, which also strongly increases the complexity of the problem. So, for the time being, research and innovation will be needed to come up with solutions for such problems as well.

In conclusion, graphics has been adopted already on a large scale to communicate and present abstract data, which shows that its value has been acknowledged, and we expect that due to the increase in size and complexity of data available, the need for more powerful and effective information visualization methods and techniques will only grow.

6   Conclusion

In this paper we have described the challenges in identifying and communicating the value of InfoVis. We have cited and posed a number of answers to the questions, "How and why is InfoVis useful?" Hopefully, the examples shown in the paper provide convincing arguments about InfoVis' value as an analytic tool. Ultimately, however, we believe that it is up to the community of InfoVis researchers and practitioners to create techniques and systems that clearly illustrate the value of the field. When someone has an InfoVis system that they use in meaningful and important ways, this person likely will not need to be convinced of the value of InfoVis.

References

1. Anscombe, F.: Graphs in statistical analysis. American Statistician 27(1), 17–21 (1973)
2. Card, S.K., Mackinlay, J., Shneiderman, B. (eds.): Readings in Information Visualization – Using Vision to Think. Morgan Kaufmann, San Francisco (1998)
3. Chincor, N., Lewis, D., Hirschman, L.: Evaluating message understanding systems: An analysis of the third message understanding conference (MUC-3). Computational Linguistics 19(3), 409–449 (1993)
4. Gilbert, E.W.: Pioneer Maps of Health and Disease in England. Geographical Journal 124, 172–183 (1958)
5. Johnson, B., Shneiderman, B.: Tree-maps: a space-filling approach to the visualization of hierarchical information structures. In: VIS '91: Proceedings of the 2nd conference on Visualization '91, Los Alamitos, CA, USA, pp. 284–291. IEEE Computer Society Press, Los Alamitos (1991)
6. Keim, D., Andrienko, G., Fekete, J.-D., Görg, C., Kohlhammer, J., Melançon, G.: Visual Analytics: Definition, Process, and Challenges. In: Kerren, A., Stasko, J.T., Fekete, J.-D., North, C.J. (eds.) Information Visualization. LNCS, vol. 4950, Springer, Heidelberg (2008)
7. Kerren, A., Stasko, J.T., Fekete, J.-D., North, C.J. (eds.): Information Visualization. LNCS, vol. 4950. Springer, Heidelberg (2008)
8. Koedinger, K.R., Anderson, J.R.: Abstract planning and perceptual chunks: Elements of expertise in geometry. Cognitive Science 14(4), 511–550 (1990)
9. Koffa, K.: Principles of Gestalt Psychology. Routledge & Kegan Paul Ltd, London (1935)
10. Larkin, J.H., Simon, H.A.: Why a diagram is (sometimes) worth 10,000 words. Cognitive Science 11, 65–100 (1987)
11. Lin, X.: Map displays for information retrieval. Journal of the American Society for Information Science 48(1), 40–54 (1997)
12. Marey, E.: La Méthode Graphique. Paris (1885)
13. McLachlan, P., Munzner, T., Koutsofios, E., North, S.: Liverac: Interactive visual exploration of system management time-series data. In: SIGCHI Conference on Human Factors in Computing Systems (CHI 2008), ACM Press, New York (2008)
14. Noack, A.: Energy-based clustering of graphs with nonuniform degrees. In: Healy, P., Nikolov, N.S. (eds.) GD 2005. LNCS, vol. 3843, pp. 309–320. Springer, Heidelberg (2006)
15. Norman, D.A.: Things That Make Us Smart: Defending Human Attributes in the Age of the Machine. Addison-Wesley Longman Publishing Co., Inc., Boston (1993)
16. Perer, A., Shneiderman, B.: Integrating statistics and visualization: Case studies of gaining clarity during exploratory data analysis. In: SIGCHI Conference on Human Factors in Computing Systems (CHI 2008), ACM Press, New York (2008)
17. Plaisant, C.: The challenge of information visualization evaluation. In: AVI '04: Proceedings of the working conference on Advanced visual interfaces, pp. 109–116. ACM Press, New York (2004)
18. Popper, K.R.: The Logic of Scientific Discovery. Basic Books, New York (1959)
19. Pousman, Z., Stasko, J., Mateas, M.: Casual information visualization: Depictions of data in everyday life. IEEE Transactions on Visualization and Computer Graphics 13(6), 1145–1152 (2007)

20. Saraiya, P., North, C., Lam, V., Duca, K.: An insight-based longitudinal study of visual analytics. IEEE Transactions on Visualization and Computer Graphics 12(6), 1511–1522 (2006)
21. Shannon, C.E., Weaver, W.: A Mathematical Theory of Communication. University of Illinois Press, Champaign (1963)
22. Spence, I., Garrison, R.F.: A remarkable scatterplot. The American Statistician, 12–19 (1993)
23. Triesman, A.: Preattentive processing in vision. Computer Vision, Graphics, and Image Processing 31(2), 156–177 (1985)
24. Tufte, E.R.: The Visual Display of Quantitative Information. Graphics Press, Cheshire (1983)
25. Tufte, E.R.: Envisioning Information. Graphics Press, Cheshire (1990)
26. Tufte, E.R.: Visual Explanations: Images and Quantities, Evidence and Narrative. Graphics Press, Cheshire (1997)
27. van Wijk, J.J.: The value of visualization. In: Proceedings IEEE Visualization 2005, pp. 79–86 (2005)
28. van Wijk, J.J., van de Wetering, H.: Cushion treemaps. In: Proceedings 1999 IEEE Symposium on Information Visualization (InfoVis'99), pp. 73–78. IEEE Computer Society Press, Los Alamitos (1999)
29. Voorhees, E., Harman, D.: Overview of the sixth Text Retrieval Conference. Information Processing and Management 36(1), 3–35 (2000)
30. Ware, C.: Information Visualization: Perception for Design. Morgan Kaufmann Publishers Inc., San Francisco (2004)
31. Wattenberg, M., Kriss, J.: Designing for social data analysis. IEEE Transactions on Visualization and Computer Graphics 12(4), 549–557 (2006)
32. Williamson, C., Shneiderman, B.: The dynamic homefinder: evaluating dynamic queries in a real-estate information exploration system. In: SIGIR '92: Proceedings of the 15th annual international ACM SIGIR conference on Research and development in information retrieval, pp. 338–346. ACM Press, New York (1992)

Evaluating Information Visualizations

Sheelagh Carpendale

Department of Computer Science, University of Calgary,
2500 University Dr. NW, Calgary, AB, Canada T2N 1N4
sheelagh@ucalgary.ca

1   Introduction

Information visualization research is becoming more established, and as a result, it is becoming increasingly important that research in this field is validated. With the general increase in information visualization research there has also been an increase, albeit disproportionately small, in the amount of empirical work directly focused on information visualization. The purpose of this paper is to increase awareness of empirical research in general, of its relationship to information visualization in particular; to emphasize its importance; and to encourage thoughtful application of a greater variety of evaluative research methodologies in information visualization.

One reason that it may be important to discuss the evaluation of information visualization, in general, is that it has been suggested that current evaluations are not convincing enough to encourage widespread adoption of information visualization tools [57]. Reasons given include that information visualizations are often evaluated using small datasets, with university student participants, and using simple tasks. To encourage interest by potential adopters, information visualizations need to be tested with real users, real tasks, and also with large and complex datasets. For instance, it is not sufficient to know that an information visualization is usable with 100 data items if 20,000 is more likely to be the real-world case. Running evaluations with full data sets, domain specific tasks, and domain experts as participants will help develop much more concrete and realistic evidence of the effectiveness of a given information visualization. However, choosing such a realistic setting will make it difficult to get a large enough participant sample, to control for extraneous variables, or to get precise measurements. This makes it difficult to make definite statements or generalize from the results. Rather than looking to a single methodology to provide an answer, it will probably will take a variety of evaluative methodologies that together may start to approach the kind of answers sought.

The paper is organized as follows. Section 2 discusses the challenges in evaluating information visualizations. Section 3 outlines different types of evaluations and discusses the advantages and disadvantages of different empirical methodologies and the trade-offs among them. Section 4 focuses on empirical laboratory experiments and the generation of quantitative results. Section 5 discusses qualitative approaches and the different kinds of advantages offered by pursuing this type of empirical research. Section 6 concludes the paper.

A. Kerren et al. (Eds.): Information Visualization, LNCS 4950, pp. 19–45, 2008.

2   Challenges in Evaluating Information Visualizations

Much has already been written about the challenges facing empirical research in information visualization [2, 12, 53, 57]. Many of these challenges are common to all empirical research. For example, in all empirical research it is difficult to pick the right focus and to ask the right questions. Given interesting questions, it is difficult to choose the right methodology and to be sufficiently rigorous in procedure and data collection. Given all of the above, appropriate data analysis is still difficult and perhaps most difficult of all is relating a new set of results to previous research and to existing theory. However, information visualization research is not alone in these difficulties; the many other research fields that also face these challenges can offer a wealth of pertinent experience and advice.

In particular, empirical research in information visualization relates to human computer interaction (HCI) empirical research, perceptual psychology empirical research, and cognitive reasoning empirical research. The relationship to empirical research in HCI is evident in that many of the tasks of interest are interface interaction tasks, such as zooming, filtering, and accessing data details [66]. The aspects of these interactive tasks that provide access to the visual representation and its underlying dataset often relate to the usability of a system. Other challenges that are shared with HCI empirical research include the difficulty of obtaining an appropriate sample of participants. If the visualization is intended for domain experts it can be hard to obtain their time. Also, when evaluating complex visualization software, it may not be clear whether the results are due to a particular underlying technique or the overall system solution. If an existing piece of software is to be used as a benchmark against which to compare an interactive visualization technique, it is likely that participants may be much more familiar with the existing software and that this may skew the results. This problem becomes more extreme the more novel a given visualization technique is. Research prototypes are not normally as smooth to operate as well established software, creating further possibilities for affecting study results and leading to controversy about testing research prototypes against the best competitive solution. Greenberg and Buxton [27] discuss this problem in terms of interaction sketches, encouraging caution when thinking about conducting usability testing on new ideas and new interface sketches in order to avoid interfering with the development of new ideas. In addition, research software does not often reach a stage in which it can support a full set of possible tasks or be fully deployable in real-world scenarios [57].

In addition to usability questions, perceptual and comprehensibility questions such as those considered in perceptual psychology are important in assessing the appropriateness of a representational encoding and the readability of visuals [30, 79]. Also, in information visualization, there are a great variety of cognitive reasoning tasks that vary with data type and character, from low-level detailed tasks to complex high-level tasks. Some of these tasks are not clearly defined, particularly those that hold some aspect of gaining new insight into the data, and may be more challenging to test empirically. Examples of low-level detailed tasks include such tasks as compare, contrast, associate, distinguish, rank, cluster, correlate, or categorize [57]; higher-level and more complex cognitive tasks include developing an understanding of data trends, uncertainties, causal relationships, predicting the future, or learning a domain [1]. Many important tasks can require weeks or months to complete. The success of information visualization is often an interplay between an expert's meta-knowledge and knowledge of other sources as well as information from the visualization in use.

While all of the above are important, a question that lies at the heart of the success of a given information visualization is whether it sheds light on or promotes insight into the data [55, 63]. Often, the information processing and analysis tasks are complex and ill-defined, such as discovering the unexpected, and are often long term or on-going. What exactly insight is probably varies from person to person and instance to instance; thus it is hard to define, and consequently hard to measure. Plaisant [57] describes this challenge as "answering questions you didn't know you had." While it is possible to ask participants what they have learned about a dataset after use of an information visualization tool, it strongly depends on the participants' motivation, their previous knowledge about the domain, and their interest in the dataset [55, 63]. Development of insight is difficult to measure because in a realistic work setting it is not always possible to trace whether a successful discovery was made through the use of an information visualization since many factors might have played a role in the discovery. Insight is also temporally elusive in that insight triggered by a given visualization may occur hours, days, or even weeks after the actual interaction with the visualization. In addition, these information processing tasks frequently involve teamwork and include social factors, political considerations and external pressures such as in emergency response scenarios. However, there are other fields of research that are also grappling with doing empirical research in complex situations. In particular, ecologists are faced with conducting research towards increasing our understanding of complex adaptive systems. Considering the defining factors of complex adaptive systems may help to shed some light on the difficulties facing empirical research in information visualization. These factors include non-linearity, holoarchy and internal causality [37, 49]. When a system is non-linear, the system behaviour comes only from the whole system. That is, the system can not be understood by decomposing it into its component parts which are then reunited in some definitive way. When a system is holoarchical it is composed of holons which are both a whole and a part. That is, the system is mutually inter-nested. While it is not yet common to discuss information analysis processes in terms of mutual nesting, in practice many information analysis processes are mutually nested. For instance, consider the processes of search and verification: when in the midst of searching, one may well stop to verify a find; and during verification of a set of results, one may well need to revert to search again. Internal causality indicates that the system is self-organizing and can be characterized by goals, positive and negative feedback, emergent properties and surprise. Considering that it is likely that a team of information workers using a suite of visualization and other software tools is some type of complex adaptive system suggests that more holistic approaches to evaluation may be needed.

Already from this brief overview, one can see that useful research advice on the evaluation of information visualization can be gathered from perceptual psychology, cognitive reasoning research, as well as human computer interaction research. Many, but not enough, information visualization researchers are already actively engaged in this pursuit. The purpose of this paper is to applaud them, to encourage more such research, and to suggest that the research community to be more welcoming of a greater variety of these types of research results.

3   Choosing an Evaluation Approach

A recent call for papers from the information visualization workshop, Beyond Time and Errors (BELIV06) held at Advanced Visual Interfaces 2006, stated that "*Controlled experiments remain the workhorse of evaluation but there is a growing sense that information visualization systems need new methods of evaluation, from longitudinal field studies, insight based evaluation and other metrics adapted to the perceptual aspects of visualization as well as the exploratory nature of discovery*" [7]. The purpose of this section is to encourage people to consider more broadly what might be the most appropriate research methods for their purposes. To further this purpose a variety of types of empirical research that can be usefully conducted are briefly outlined and these differing types are discussed in terms of their strengths and weaknesses. This discussion draws heavily from McGrath's paper Methodology Matters [50] that was initially written for social scientists. However, while social scientists work towards understanding humans as individuals, groups, societies and cultures, in information visualization – similarly to HCI – we are looking to learn about how information visualizations do or do not support people in their information tasks and/or how people conduct their information related tasks so that visualization can be better designed to support them. To gain this understanding we sometimes study people using information visualization software and sometimes it may be important to study people independently of that software, to better understand the processes we are trying to support.

There are some commonalities to all studies. They all must start with some question or questions that will benefit from further study. Also, they all must relate their research questions to the realm of existing ideas, theories and findings. These ideas, theories, and concepts are needed to relate the new study to existing research. For example, the results from a new study might be in contrast to existing ideas, in agreement with existing ideas, or offer an extension of or variation to existing ideas. A study must also have a method. This is what this section is about – possible types of empirical methodologies.

All methods offer both advantages and disadvantages. One important part of empirical research is choosing the most appropriate research methods for your content, your ideas, and your situation. The fact that methods both provide and limit evidence suggests that making use of a wide variety of methodologies will, in time, strengthen our understandings. Thus, both conducting a greater variety of studies and encouraging this by publishing research that employs a greater variety of methodologies will help to develop a better understanding of the value of information visualization and its potential in our communities.

When conducting a study there are three particularly desirable factors: generalizability, precision, and realism [50]. Ideally, one would like all of these factors in one's results. However, existing methodologies do not support the actualization of all three simultaneously. Each methodology favours one or two of these factors, often at the expense of the others; therefore the choice of a methodology for a particular goal is important. To define these terms (as used in McGrath [50]):

- **Generalizability**: a result is generalizable to the extent to which it can apply to other people (than those directly in the study) and perhaps even extend to other situations.

- **Precision:** a result is precise to the degree to which one can be definite about the measurements that were taken and about the control of the factors that were not intended to be studied.
- **Realism:** a result is considered realistic to the extent to which the context in which it was studied is like the context in which it will be used.

Figure 1 (adapted and simplified from McGrath [50]) shows the span of common methodologies currently in practice in the social sciences. They are positioned around the circle according to the labels: most precision, most generalizability and most realism. The closer a methodology is placed to a particular label, the more that label applies to that methodology. Next, these methodologies are briefly described. For fuller descriptions see McGrath 1995.

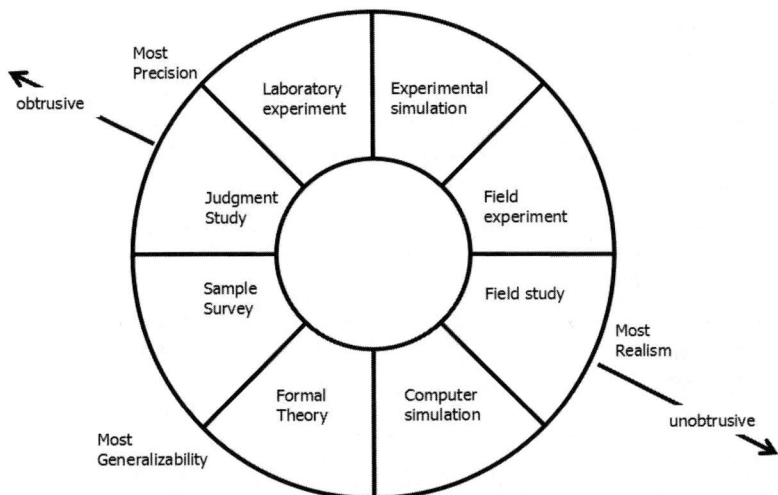

Fig. 1. Types of methodologies organized to show relationships to precision, generalizability and realism. (adapted, simplified from McGrath 1995)

Field Study: A field study is typically conducted in the actual situation, and the observer tries as much as possible to be unobtrusive. That is, the ideal is that the presence of the observer does not affect what is being observed. While one can put considerable effort into minimizing the impact of the presence of an observer, this is not completely possible [50]. Examples of this type of research include ethnographic work in cultural anthropology, field studies in sociology, and case studies in industry. In this type of study the realism is high but the results are not particularly precise and likely not particularly generalizable. These studies typically generate a focused but rich description of the situation being studied.

Field Experiment: A field experiment is usually also conducted in a realistic setting; however, an experimenter trades some degree of unobtrusiveness in order to obtain more precision in observations. For instance, the experimenter may ask the participants to perform a specific task while the experimenter is present. While realism is still high, it has been reduced slightly by experimental manipulation. However, the necessity of long observations may be shortened and results may be more readily interpretable and specific questions are more likely to be answered.

Laboratory Experiment: In a laboratory experiment the experimenters fully design the study. They establish what the setting will be, how the study will be conducted, what tasks the participants will do, and thus plan the whole study procedure. Then the experimenter gets people to participate as fully as possible following the rules of the procedure within the set situation. Carefully done, this can provide for considerable precision. In addition, non-realistic behaviour that provides the experimenter more information can be requested such as a 'think aloud' protocol [43]. Behaviour can be measured, partly because it is reasonably well known when and where the behaviour of interest may happen. However, realism is largely lost and the degree to which the experimenter introduces aspects of realism will likely reduce the possible precision.

Experimental Simulation: With an experimental simulation the experimenter tries to keep as much of the precision as possible while introducing some realism via simulation. There are examples where this approach is essential such as studying driving while using a cell phone or under some substance's influence by using a driving simulator. Use of simulation can avoid risky or un-ethical situations. Similarly although less dramatically, non-existent computer programs can be studied using the 'Wizard of Oz' approach in which a hidden experimenter simulates a computer program. This type of study can provide us with considerable information while reducing the dangers and costs of a more realistic experiment.

Judgment Study: In a judgment study the purpose is to gather a person's response to a set of stimuli in a situation where the setting is made irrelevant. Much attention is paid to creating 'neutral conditions'. Ideally, the environment would not affect the result. Perceptual studies often use this approach. Examples of this type of research include the series of studies that examine what types of surface textures best support the perception of 3D shape (e.g. [34, 38]), and the earlier related work about the perception of shape from shading [39]. However, in assessing information visualizations this idea of setting a study in neutral conditions must be considered carefully, as witnessed by Reilly and Inkpen's [62] study which showed that the necessity for an interactive technique developed to support a person's mental model during transition from viewing one map to another (subway map to surface map) was dependent on the distractions in the setting. This transition technique relates to ideas of morphing and distortion in that aspects of the map remain visible while shifting. These studies in a more neutral experiment setting showed little benefit, while the same tasks in a noisy, distracting setting showed considerable benefit.

Sample Survey: In a sample survey the experimenter is interested in discovering relationships between a set of variables in a given population. Examples of these types of questions include: of those people who discover web information visualization tools how many return frequently and are their activities social or work related? Of those people who have information visualization software available at work what is the frequency of use? Considering the increased examples of information visualization results and software on the web, is the general population's awareness of and/or use of information visualization increasing? In these types of studies proper sampling of the population can lead to considerable generalizability. However, while choosing the population carefully is extremely important, often it is difficult to control. For example in a web-based survey, all returned answers are from those types of people who are willing to take the time, fill out the questionnaire, etc. This is a type of bias and thus reduces generalizability. Also, responses are hard to calibrate. For instance, a particular paper reviewer may never give high scores and the meta-reviewer may know this and calibrate accordingly or may not know this. Despite these difficulties, much useful information can be gathered this way. We as a community must simply be aware of the caveats involved.

Formal Theory: Formal theory is not a separate experimental methodology but an important aspect of all empirical research that can easily be overlooked. As such, it does not involve the gathering of new empirical evidence and as a result is low in both precision and realism. Here, existing empirical evidence is examined to consider the theoretical implications. For example, the results of several studies can be considered as a whole to provide a higher-level or meta-understanding or the results can be considered in light of existing theories to extend, adjust or refute them. Currently this type of research is particularly difficult to publish in that there are no new information visualizations and no new empirical results. Instead, the contribution moves towards the development of theories about the use of and practicality of information visualizations.

Computer Simulation: It is also possible to develop a computer simulation that has been designed as logically complete. This method is used in battle simulation, research and rescue simulation, etc. This type of strategy can be used to assess some visualizations. For instance, a visualization of landscape vegetation that includes models of plant growth and models of fire starts and spread can be set to simulate passage of several hundred years. If the resulting vegetation patterns are comparable to existing satellite imagery this provides considerable support for the usefulness of models [22]. Since this type of research strategy does not involve participants, discussion of generalizability over populations is not applicable. Also, since the models are by definition incomplete, notions of precision in measurement are often replaced with stochastic results. On the other hand it does provide a method of validation and offers a parallel with which we can study realistic situations, such as explosions, turbulence in wind tunnels, etc.

4   Focus on Quantitative Evaluation

Quantitative evaluations, most well known as laboratory experiments or studies, are those methodologies in which precision is relatively high and in which some declaration can be made about the possible generalization to a larger population. These declarations can include information about the characterization of this larger population and how likely it is that the generalization will hold. These types of experiments or studies are part of the traditional empirical scientific experimental approach and have evolved and been refined through the centuries of scientific research. Science does and has depended on these methods. Slowly, through careful and rigorous application of the experimental process, knowledge has been built up, usually one piece at a time.

The experiments or studies involve a rigorous process of hypothesis development, identification and control of the independent variables, observation and measurement of the dependent variables, and application of statistics which enable the declaration of the confidence with which the results can be taken. In these formal studies or controlled evaluations, the experimenter controls the environment or setting, manipulates chosen factor(s) or variable(s) – the independent variable(s) - in order to be able to measure and observe the affect this manipulation has on one or more other factors – the dependent variable(s). Ideally no other factors change during the experiment. Once the changes to the dependent variables have been measured, statistical methods can be applied to understand the relative importance of the results. Done with sufficient thoroughness, this process can arrive at facts about which we can be relatively certain. The application of this scientific process will to try to reduce the overall complexity by fine tuning particular questions or hypotheses, using these hypotheses to allow one to cull some of the complexity by trying to eliminate as many of the extraneous variables as possible. Traditionally experiments of this type are used to shed light on cause and effect relationships; that is, to discover whether changes in some factor result in changes to another factor.

This idea that we can observe simpler, more manageable subsets of the full complex process is appealing, and it is clear from centuries of experiments that much can be learnt in this manner.

4.1   Quantitative Methodology

Since quantitative empirical evaluations have evolved over the centuries the methodology has become relatively established (Figure 2). This brief overview is included for completeness; the interested reader should refer to the many good books on this subject [15, 17, 33]. This methodology includes:

- **Hypothesis Development:** Much of the success of a study depends on asking an interesting and relevant question. This question should ideally be of interest to the broader research community, and hopefully answering it will lead to a deeper or new understanding of open research questions. Commonly the importance of the study findings results from a well thought through hypothesis, and formulating this question precisely will help the development of the study.

- **Identification of the Independent Variables:** The independent variables are the factors to be studied which may (or may not) affect the hypothesis. Ideally the number of independent variables is kept low to provide more clarity and precision in the results.

- **Control of the Independent Variables:** In designing the experiment the experimenter decides the manner in which the independent variables will be changed.

- **Elimination of Complexity:** In order to be clear that it is actually the change in the independent variable that caused the study's result, it is often the case that other factors in the environment need to be controlled.

- **Measurement of the Dependent Variables:** Observations and measurements are focused on the dependent variables as they change or do not change in

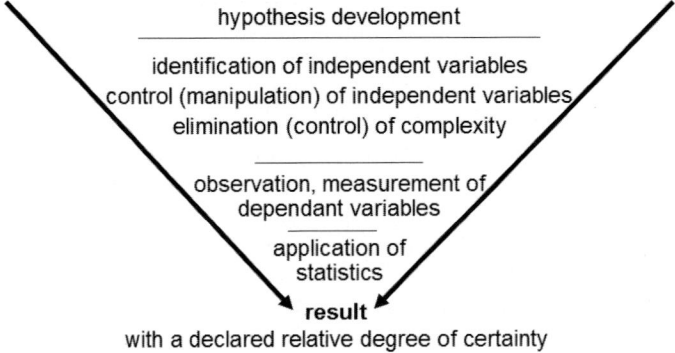

hypothesis development

identification of independent variables
control (manipulation) of independent variables
elimination (control) of complexity

observation, measurement of
dependant variables

application of
statistics

result
with a declared relative degree of certainty

Fig. 2. A simple schematic of the traditional experimental process.

response to the manipulation of the independent variable. The aspects to be measured are often called metrics. Common metrics include: speed, accuracy, error rate, satisfaction, etc.

- **Application of Statistics:** The results collected can then be analysed through the application of appropriate statistics. It is important to remember that statistics tell us how sure we can be that these results could (or could not) have happened by chance. This gives a result with a relative degree of certainty. There are many good references such as Huck [33].

These steps sound deceptively simple but doing them well requires careful and rigorous work. For instance, it is important that the study participants are valued, that they are not over-stressed, and that they are given appropriate breaks, etc. Also, exactly what they are being asked to do must be clear and consistent across all participants in your study. Since small inconsistencies such as changes in the order of the instructions can affect the results, the common recommendation is that one scripts the explanations. Perhaps most importantly, to eliminate surprises and work out the details, it is best to pilot – run through the experiment in full – repeatedly.

4.2   Quantitative Challenges

Even though these types of experiments have been long and effectively used across all branches of science, there remain many challenges to conducting a useful study. We mention different types of commonly-discussed errors and validity concerns and relate these to the McGrath's discussion as outlined in Section 3. In this discussion we will use a simple, abstract example of an experiment that looks at the effect of two visualization techniques, VisA and VisB, on performance in search. There are several widely discussed issues that can interfere with the validity of a study.

Conclusion Validity: Is there a relationship? This concept asks whether within the study there is a relationship between the independent and the dependent variables. Important factors in conclusion validity are finding a relationship when one does not exist (type I error) and not finding a relationship when one does exist (type II error).

Table 1. Type I and Type II Errors

		Reality	
		H_0 TRUE	H_0 FALSE
Experimental decision	H_0 TRUE	ok	Type II
	H_0 FALSE	Type I	ok

Type I and Type II Errors: If one is interested in which visualization technique VisA or VisB helps people conduct a particular task faster one might formulate a null hypothesis (H_0) – *there is no difference in speed of search between VisA and VisB*. The possible type I, false negative, and type II, false positive, errors are specified in Table 1. The columns represent whether the null hypothesis is true or false in reality and the rows show the decision made based on the results of the experiment. Ideally the results of the experiment reflect reality and that if the hypothesis is false (or true) in reality it will show as false (or true) in the experiment. However, it is possible that the hypothesis is true in reality – VisA does support faster search than VisB – but that this fails to be revealed by the experiment. This is a type II error. A type I error occurs if the null hypothesis is true in reality (there is no difference) and one concludes that there *is* a difference. Type I errors are considered more serious. That is, it is considered worse to claim that VisA improves search when it does not, than to say there was no measurable difference.

Internal Validity: Is the relationship causal? This concept is important when an experiment is intended to reveal something about causal relationships. Thus, internal validity will be important in our simple example because the study is looking at what effect VisA and VisB have on search. The key issue here is whether the results of one's study can properly be attributed to what happened within the experiment. That is, that no other factors influenced or contributed to the results seen in the study. Another way of asking this question is: are there possible alternate causes for the results seen in the study?

Construct Validity: Can we generalize to the constructs (ideas) the study is based on? This concept considers whether the experiment has been designed and run in a manner that answers the intended questions. This is an issue about whether the right factors are being measured or whether the factors the experimenter intends to measure are actually those being measured. For instance, does the experiment measure the difference due to the techniques VisA and VisB or the difference in participant's familiarity with VisA and VisB. For instance, if the construct is that a person will have higher satisfaction when using VisB, does measuring error rates and completion times provide answers for this construct? An important part of this concept of construct validity is **measurement validity**. Measurement validity is concerned with questions such as is one measuring what one intends to measure and is the method of measurement reliable and consistent. That is, will the same measurement process provide the same results when repeated?

External Validity: Can we generalize the study results to other people/places/ times? External validity is concerned with the extent to which the result of a study can be generalized. If a study has good internal and construct validity the results apply to the setting, time, and participants being studied. The extent to which the results apply beyond the immediate setting, time and participants depends, for participants, on the participant sample and the population from which it was drawn. For instance, in practice it is common to draw participants from the geographic region in which the study is run. Does this mean that the results only apply to people from that region? If culture has a possible impact on the results, they may not generalize. If one addresses the need to include cultural variation by recruiting participants from different cultures from a university's foreign students, one might have at least partially addressed the need to run the study across cultural variations but now have limited the demographic to university students which may introduce its own skew. Understanding the population to which one would like to be able to generalize the study results and successfully obtaining an appropriate participant sample is a difficult issue. This does not mean we can not learn from more specific participant samples. It does mean that reporting the demographics of the sample and being cautious about generalizations is important. Participant sample choice is just one factor influencing external validity. Setting includes other factors such as noise, interruption, and distractions. Possible temporal factors include events that occurred before or are anticipated after the experiment.

Ecological Validity: Ecological validity discussions focus on the degree to which the experimental situation reflects the type of environment in which the results will be applied. This concept relates strongly to McGrath's concept of realism. It is distinct from the idea of external validity, in that external validity is concerned with whether the experimental results generalize to other situations, while ecological validity is concerned with how closely the experimental settings matches the real setting in which the results might be applied. Thus it is possible to have good ecological validity; the study is conducted on site, but that the results are applicable only to that site. This would indicate poor external validity in that the results do not generalize beyond the specific setting.

4.3   Quantitative Studies Summary Remarks

The number of quantitative studies in information visualization is increasing. Early examples include the series of studies done by Purchase and her collaborators that examine the impact of graph drawing aesthetics on comprehension and usability [58, 59, 60, 61]. Dumais et al. [16] explored use of context techniques in web search. Forlines et al. [23] looked at the effect of display configuration on relationship between visual search and information visualization tasks. Recently, Willet et al. [81] studied embedding information visualizations in widgets.

Quantitative experiments have formed the backbone of experimental science and it is to be expected that they will continue to do so. However, it is relatively easy to find fault in any given experiment because all factors can not usually be completely controlled. If they are completely controlled, external and ecological validity can be impacted. This is particularly true for studies involving humans. Designing and working with experiments is often a matter of making choices about what factors are important and understanding the strengths and limitations of any given study and its results. As a community it is important that we recognise that we are working towards a larger understanding and that any given study will not present the bigger answer. It instead will contribute to a gradual building of a bigger understanding. For this bigger understanding we need to encourage authors to openly discuss the limitations of their studies, because both the results and the limitations are important. This is also true for negative results. It can be just as important to understand when there are no differences among techniques and when these differences exist.

5   Focus on Qualitative Evaluation

Qualitative inquiry works toward achieving a richer understanding by using a more holistic approach that considers the interplay among factors that influence visualizations, their development, and their use [56]. Qualitative techniques lend themselves to being more grounded in more realistic settings and can also be incorporated into all types of studies. This includes qualitative studies conducted as part of the design process [64, 73], in situ interviews [83], field studies [72], and use of observational studies to create design and evaluative criteria that are derived from observed data [71]. These types of studies offer potential for improved understanding of existing practices, analysis environments, and cognitive task constraints as they occur in real or realistic settings. In providing a brief overview of a variety of qualitative methods, we hope to spark further research and application of qualitative methods in information visualization; to expand our empirical approaches to include the application of qualitative methods to design and evaluation; and to encourage a wider acceptance of these types of research methodologies in our field.

5.1   Qualitative Methods

At the heart of qualitative methods is the skill and sensitivity with which data is gathered. Whether the records of the data gathered are collected as field notes, artefacts, video tapes, audio tapes, computer records and logs, or all of these, in qualitative empirical approaches there are really only two primary methods for gathering data: observations and interviews. Observation and interview records are usually kept continually as they occur, as field notes, as regular journal entries as well as often being recorded as video or audio tapes. Artefacts are collected when appropriate. These can be documents, drawings, sketches, diagrams, and other objects of use in the process being observed. These artefacts are sometimes annotated as part of use practices or in explanation. Also, since the communities we are observing are often technology users, technology-based records can also include logs, traces, screen captures, etc. Both observation and interviewing are skills and as such develop with practice and can, at least to some extent, be learnt. For full discussions on these skills there are many useful books such as Seidman [65] and Lofland and Lofland [45].

5.1.1 Observation Techniques
The following basic factors have been phrased in terms of developing observational records but implicitly also offer advice on what to observe:

- Try to keep jotting down notes unobtrusively. Ideally, notes are taken as observations occur; however, if one becomes aware that one's note taking is having an impact on the observations, consider writing notes during breaks, when shielded, or at the end of the day.

- Minimize the time gap from observations to note taking. Memory can be quite good for a few hours but does tend to drop off rapidly.

- Include in observations the setting, a description of the physical setup, the time, who is present, etc. Drawing maps of layouts and activities can be very useful.

- Remember to include both the overt and covert in activities and communications. For example, that which is communicated in body language and gestures, especially if it gets understood and acted upon, is just as important as spoken communications. But be careful of that grey area where one is not sure to what extent a communication occurred.

- Remember to include both the positive and negative. Observed frustrations and difficulties can be extremely important in developing a fuller understanding.

- Do not write notes on both sides of a paper. This may seem trivial but experienced observers say this is a must [6]. You can search for hours, passing over many times that important note that is on the back of another note.

- Be concrete whenever possible.

- Distinguish between word-for-word or verbatim accounts and those you have paraphrased and/or remembered.

5.1.2 Interview Techniques

These are a few brief points of advice about interviewing. Do remember that while sorting out the right questions to ask is important, actively listening to what the participant says is the most important of all interviewing skills.

- Make sure that you understand what they are telling you and that the descriptions, explanations they are giving you are complete enough. However, when asking for clarification, try to avoid implying that their explanations are poor because one does not want to make one's participants defensive. Ask instead for what they meant by particular word usage or if they would explain again. The use of the word *again* implies that the interviewer did not catch it all rather than the explanation was incomplete.

- Limit your inclination to talk. Allow for pauses in the conversation, sometimes note taking can be useful here. The participant will expect you to be taking notes. In this situation note taking can actually express respect for what the participant has said.

- Remember that the default is that the participant will regard the interview to some extent as public and thus will tell you the public version. Do listen for and encourage the less formal, less guarded expression of their thoughts. One example, from Seidman [65], is the use of the word 'challenge'. Challenge is an excepted term for a problem. The details of the problem might be explained more fully if one asks what is meant in the given situation by the word challenge.

- Follow up on what the participant says. Do allow the interview to be shaped by the information your participant is sharing.

- Avoid leading questions. An important part of minimizing experimenter bias is wording questions carefully so as to avoid implying any types of answers. For example, asking a participant what a given experience was like for them, leaves space for their personal explanations.

- Ask open ended questions. This can involve asking for a temporal reconstruction of an event or perhaps a working a day or asking for a subjective interpretation of an event.

- Ask for concrete details. These can help trigger memories.

- With all the above do remember that one of the most important pluses of an interview process is the humanity of interviewer. Being present, aware and sensitive to the process is your biggest asset. These guidelines are just that; guidelines to be used when useful and ignored when not.

5.2   Types of Qualitative Methodologies

This section is not intended to be a complete collection of all types of qualitative inquiry. Rather it is meant to give an overview of some of the variations possible, set in a discussion about when and where they have proven useful. This overview is divided into three sections. First, the type of qualitative methodologies often used in conjunction with or as part of more quantitative methodologies is discussed. Then, we mention the approaches taken in the area of heuristic, or, as they are sometimes referred to 'discount', inspection methodologies. The last section will cover some study methodologies that are intentionally primarily qualitative.

5.2.1 Nested Qualitative Methods

While qualitative methodologies can be at the core of some types of studies, some aspects of qualitative inquiry are used in most studies. For instance, data gathered by asking participants for their opinions or preferences is qualitative. Gorard [26] argues that quantitative methods can not ignore the qualitative factors of the social context of the study and that these factors are, of necessity, involved in developing an interpretation of the study results. There are many methods used as part of studies such as laboratory experiments that provide us with qualitative data. The following are simply a few examples to illustrate how common this mixed approach is.

Experimenter Observations: An important part of most studies is that the experimenter keeps notes of what they observe as it is happening. The observations themselves can help add some degree of realism to the data and the practice of logging these observations as they happen during the study helps make them more reliable than mere memory. However, they are experimenter observations and as such are naturally subjective. They do record occurrences that were not expected or are not measurable so that they will also form part of the experimental record. These observations can be helpful during interpretation of the results in that they may offer explanations for outliers, point towards important experimental re-design, and suggest future directions for study. Here, experimenter observations augment and enrich the primarily quantitative results of a laboratory experiment and in this they play an important but secondary role.

Think-Aloud Protocol: This technique, which involves encouraging participants to speak their thoughts as they progress through the experiment, was introduced to the human-computer-interaction community by [43]. Discussions about this protocol in psychology date back to 1980 [19, 20, 21]. Like most methodologies, this one also involves tradeoffs. While it gives the experimenter/observer the possibility of being aware of the participants' thoughts, it is not natural for most people and can make a participant feel awkward; thus, think aloud provides additional insight while also reducing the realism of the study. However, the advantage for hearing about a participant's thoughts, plans, and frustrations frequently out-weigh the disadvantages and this is a commonly used technique. Several variations have been introduced such as 'talk aloud' which asks a participant to more simply announce their actions rather than their thoughts [21].

Collecting Participant Opinions: Most laboratory experiments include some method by which participant opinions and preferences are collected. This may take the form of a simple questionnaire or perhaps semi-structured interviews. Most largely quantitative studies such as laboratory experiments do ask these types of questions, often partially quantifying the participant's response by such methods as using a Likert scale [44]. A Likert scale asks a participant to rate their attitude according to degree. For instance, instead of simply asking a participant, 'did you like it?' A Likert scale might ask the participant to choose one of a range of answers 'strongly disliked,' 'disliked,' 'neutral,' 'liked,' or 'strongly liked.'

Summary of Nested Qualitative Methods: The nested qualitative methods mentioned in this section may be commonplace to many readers. The point to be made here is that in the small, that is as part of a laboratory experiment, inclusion of some qualitative methods is not only commonplace, its value is well recognized. This type of inclusion of qualitative approaches adds insight, explanations and new questions. It also can help confirm results. For instance, if participants' opinions are in line with quantitative measures – such as the fastest techniques being the most liked – this confirms the interpretation of the fastest technique being the right one to chose. However, if they contradict – such as the fastest techniques not being preferred – interesting questions are raised including questioning the notion that fastest is always best.

5.2.2 Inspection Evaluation Methods

We include a discussion of inspection methods because, while they are not studies per se, they are useful, readily available, and relatively inexpensive evaluation approaches. The common approach is to use a set of heuristics as a method of focusing attention on important aspects of the software – interface or visualization – which need to be considered [54]. These heuristics or guidelines can be developed by experts or from the writings of experts. Ideally, such an inspection would be conducted by individual experts or even a group of experts. However, it has been shown that in practice, that a good set of heuristics can still be effective in application if a few, such as three or four, different people apply them [54]. For information visualization it is important to consider exactly what visualization aspects a given set of heuristics will shed light on.

Usability Heuristics: These heuristics, as introduced and developed by Nielson and Mack [1994], focus on the usability of the interface and are designed to be applied to any application, thus are obviously of use to information visualizations. They will help make sure that general usability issues are considered. These heuristics are distilled down to ten items – visibility of system status, match between system and real world, personal control and freedom, consistency and standards, error prevention, recognition rather than recall, flexibility and efficiency, aesthetic and minimalist design, errors handling, and help and documentation.

Collaboration Heuristics: When interfaces are designed for collaboration, two additional major categories arise in importance: communication and coordination. Baker et al. [4] developed a set of heuristics that explore these issues based on the Mechanics of Collaboration [29]. As information visualizations start to be designed for collaborative purposes, both distributed [31, 78] and co-located [35], these heuristics will also be important.

Information Visualization Heuristics: While the usability heuristics apply to all infovis software and the collaboration heuristics apply to the growing body of collaborative information visualizations, there are areas of an information visualization that these at best gloss over. In response, the Information Visualization research community has proposed a variety of specific heuristics. Some pertain to given data domains such as ambient displays [46] and multiple view visualizations [5]. Others focus on a specific cognitive level, for instance knowledge and task [1], or task and usability [66]. Tory and Möller [74] propose the use of heuristics based on both visualization guidelines and usability. As explored by Zuk and Carpendale [84], we can also consider developing heuristics based on the advice from respected experts such as design advice collected from Tufte's writings [75, 76, 77], semiotic considerations as expressed by Bertin [8] and/or research in cognitive and perceptual science as collected by Ware [79]. Alternatively, we can start from information visualization basics such as presentation, representation and interaction [68]. However, a concept such as presentation cuts across design and perception, while representation advice, such as what types of visuals might best represent what types of data, might be distilled from the guidelines put forth by Bertin [8] and from an increasing body of cognitive science as gathered in Ware [79]. Sorting out how to best condense these is a task in itself [52, 85]. "At this stage of development of heuristics for information visualization we have reached a similar problem as described by Nielson and Mack [54]. It is a difficult problem to assess which list(s) are better for what reasons and under what conditions. This leads to the challenges of developing an optimal list that comprises the most important or common Information Visualization problems" (page 55, [85]).

Summary of Inspection Evaluation Methods: While experience in the human computer interaction communities and the growing body of information visualization specific research indicates that heuristics may prove a valuable tool for improving the quality of information visualizations, there is considerable research yet to be conducted in the development of appropriate taxonomies and application processes for heuristics in information visualization.

The currently recommended application approach for usability heuristics is that evaluators apply the heuristics in a two pass method. The first pass is done to gain an overview and second is used to asses in more detail each interface component with each heuristic [54]. The original use indicated that in most situations three evaluators would be cost effective and find most usability problems [54]. However, subsequent use of heuristics for web site analysis appears to sometimes need more evaluators [9, 69]. Further, this may depend on the product. While application of heuristics has not yet been formally studied in terms of web sites, it does introduce the possibility that information visualization heuristics may also need to be data, task or purpose specific.

Heuristics are akin to the design term *guidelines* in that both provide a list of advice. Design guidelines are often usefully applied in a relatively ad hoc manner as factors to keep in mind during the design process and heuristic lists can definitely be similarly used. While there are definitely benefits that accrue in the use of guidelines and heuristics, it is important to bear in mind that they are based on what is known to be successful and thus tend not to favour the unusual and the inventive. In the design world, common advice is that while working without knowledge of guidelines is foolish, following them completely is even worse.

5.2.3 Qualitative Methods as Primary

A common reason for using qualitative inquiry is to develop a richer understanding of a situation by using a more holistic approach. Commonly, the qualitative research method's goal is to collect data that enables full, rich descriptions rather than to make statistical inferences [3, 14]. There are a wealth of qualitative research methods that can help us to gain a better understanding of the factors that influence information visualization use and design. Just as we have pointed out how qualitative methods can be effectively used within quantitative research, qualitative research can also include some quantitative results. For instance, there may be factors that can be numerically recorded. These factors can then be presented in combination with qualitative data. For example, if a questionnaire includes both fixed-choice questions and open ended questions, quantitative measurement and qualitative inquiry are being combined [56].

Qualitative methods can be used at any time in the development life cycle. A finished or near to finished product can be assessed via case studies or field studies. Also, there is a growing use of these methods as a preliminary step in the design process. The HCI and particularly the computer-supported cooperative work (CSCW) research communities have successfully been using qualitative methods to gain insight that can inform the initial design. CSCW researchers have learned a lot about how to support people working together with technology through pre-design observation and qualitative analysis of how people work together without technology. The basic idea is that through observations of participants' interactions with physical artefacts, a richer understanding of basic activities can be gained and that this understanding can be used to inform interface design. This approach generally relies on observation of people, inductive derivation of hypotheses via iterative data collection, analysis, and provisional verification [14]. For example, Tang's study of group design activities around shared workspaces revealed the importance of gestures and the workspace itself in mediating and coordinating collaborative work [73]. Similarly, Scott et al. [64] studied traditional tabletop gameplay and collaborative design, specifically focusing on the use of tabletop space, and the sharing of items on the table. Both studies are an example of how early evaluation can inform the design of digital systems. In both cases, the authors studied traditional, physical contexts first, to understand participants' interactions with the workspace, the items in the workspace, and within the group. The results of these experiments are regarded as providing important information about what group processes to support and some indication about how this might be done. This type of research can be particularly important in complex or sensitive scenarios such as health care situations [72]. Brereton and McGarry [11] observed groups of engineering students and professional designers using physical objects to prototype designs. They found that the interpretation and use of physical objects depended greatly on the context of its placement, indicating that the context of people's work is important and is difficult to capture quantitatively. Their goal was to determine implications for the design of tangible interfaces. Other examples include Saraiya et al. [63] who used domain expert assessments of insight to evaluate bioinformatics visualizations, while Mazza and Berre [48] used focus groups and semi-structured interviews in their analysis of visualization approaches to support instructors in web-based distance education.

The following are simply examples of empirical methods in which gathering of qualitative data is primary. There are many others; for instance, Moggridge [51] mentions that his group makes active use of fifty-one qualitative methods in their design processes.

In Situ Observational Studies: These studies are at the heart of field studies. Here, the experimenter gets permission to observe activities as they take place in situ. In these studies the observer does their best to remain unobtrusive during the observations. The ideal in Moggridge's terms is to become as a 'fly on the wall' that no one notices [51]. This can be hard to achieve in an actual setting. However, over time a good observer does usually fade into the background. Sometimes observations can be collected via video and audio tapes to avoid the more obvious presence of a person as observer but sometimes making such recordings is not appropriate as in medical situations. In these studies the intention is usually to gather a rich description of the situation being observed. However, there is both a difference and an overlap in the type of observations to be gathered when the intention is (a) to better understand the particular activities in a given of setting, or (b) to use these observations to inform technology design. Thus, because different details are of prime interest it is important that our research community conducts these types of observational studies to better inform initial design as well as to better understand the effectiveness of new technology in use. These studies have high realism, result in rich context explicit data and are time and labour intensive when it comes to both data collection and data analysis.

Participatory Observation: This practice is the opposite of participatory design. Here an information visualization expert becomes part of the application expert's team to experience the work practices first hand rather than application experts becoming part of the information visualization design team. In participatory observation, additional insights can be gained through first-hand observer experience of the tasks and processes of interest in the context of the real world situation. Here, rather than endeavouring to be unobtrusive, the observer works towards becoming an accepted part of the community. Participatory observation is demonstrably an effective approach since as trust and rapport develop, an increasingly in-depth understanding is possible. Our research community is interested in being able to better understand the work practices of many different types of knowledge workers. These workers are usually highly trained, highly paid, and often under considerable time pressures. Not surprisingly, they are seldom willing to accept an untrained observer as part of their team. Since information visualization researchers are of necessity highly trained themselves, it is rare that an information visualization researcher will have the necessary additional training to become accepted as a participatory observer. However, domain expertise is not always essential for successful participatory observation. Expert study participants can train an observer on typical data analysis tasks – a process which may take several hours, and then "put them to work" on data analysis using their existing tools and techniques. The observer keeps a journal of the experience and the outcomes of the analysis were reviewed with the domain experts for validity. Even as a peripheral participant, valuable understandings of domain, tasks, and work culture can be developed which help clarify values and assumptions about data, visualizations, decision making and data insights important to the application domain. These understandings and constructs can be important to the information visualization community in the development of realistic tools.

Laboratory Observational Studies: These studies use observational methodologies in a laboratory setting. A disadvantage of in situ observations is that they often require lengthy observations. For instance, if the observer is interested in how an analyst uses visual data, they will have to wait patiently until the analyst does this task. Since an analyst may have many other tasks – meetings, conference calls, reports, etc. – this may take hours or even days. One alternative to the lengthy in situation wait is to design an observational experiment in which, similarly to a laboratory experiment, the experimenter designs a setting, a procedure and perhaps even a set of tasks. Consider, for example, developing information visualizations to support co-located collaboration. Some design advice on co-located collaborative aspects is available in the computer supported cooperative work literature [35]. However, while this advice is useful, it does not inform us specifically about how teams engage in collaborative tasks when using visual information. Details such as how and when visualizations will be shared and what types of analysis processes need to be specifically supported in collaborative information visualization systems were missing. Here, an observational approach is appropriate because the purpose is to better understand the flow and nature of the collaboration among participants, rather than answering quantifiable lower-level questions. In order to avoid temporal biases in existing software, pencil and paper based visualizations were used. This allowed for the observation of free arrangement of data, annotation practices, and collaborative processes unconstrained by any particular visualization software [36].

Contextual Interviews: As noted in Section 5.1, interviewing in itself is core to qualitative research. Conducting an interview about a task, setting, or application of interest within the context in which this work usually takes place is just one method that can enrich the interview process. Here the realism of the setting helps provide the context that can bring to mind the day-to-day realities during the interview process (for further discussion see Holtzblatt and Beyer 1998). For example, to study how best to support the challenging problem of medical diagnosis, observing and interviewing physicians in their current work environment might help to provide insights into their thought processes that would be difficult to capture with other methodologies. A major benefit of qualitative study can be seeing the big picture – the context in which a new visualization support may be used. The participants' motives, misgivings, and opinions shed light on how they relate to existing support, and can effectively guide the development of new support. This type of knowledge can be very important at the early stage of determining what types of information visualizations may be of value.

Summary of qualitative methods as primary: These four methods are just examples of a huge variety of possibilities. Other methods include action research [42], focus groups [48], and many more. All these types of qualitative methods have the potential to lessen the task and data comprehension divide between ourselves as visualization experts and the domain experts for whom we are creating visualizations. That is, while we can not become analysts, doctors, or linguists, we can gain a deeper understanding of how they work and think. These methods can open up the design space, revealing new possibilities for information visualizations, as well as additional criteria on which to measure success.

5.3   Challenges for Qualitative Methods

A considerable challenge to qualitative methods is that they are particularly labour intensive. Gathering data is a slow process and rich note taking is an intensive undertaking, as are transcribing and subsequent analysis.

5.3.1 Sample Sizes

Sample sizes for qualitative research are determined differently than for quantitative research. Since qualitative research is not concerned with making statistically significant statements about a phenomenon, the sample sizes are often lower than required for quantitative research. Often, sample sizes are determined during the study. For instance, a qualitative inquiry may be continued until one no longer appears to be gaining new data through observation [3]. There is no guideline to say when this 'saturation' may occur [70]. Sample sizes may vary greatly depending on the scope of the research problem but also the experience of the investigator. An experienced investigator may reach a theoretical saturation earlier than a novice investigator. Also, because each interview and/or observation can result in a large amount of data, sometimes compromises in sample size have to be made due to considerations about the amount of data that can be effectively processed.

5.3.2 Subjectivity

Experimenter subjectivity can be seen as an asset because of the sensitivity that can be brought to the observation process. The quality of the data gathering and analysis is dependent on the experience of the investigator [56]. However, the process of gathering any data must be concerned with obtaining representative data. The questions circle about whether the observer has heard or understood fully and whether these observations are reported accurately. Considerations include:

- Is this a first person direct report? Otherwise normal common sense about 2nd, 3rd, and 4th hand reports needs to be considered.

- Does the spatial location of the observer provide an adequate vantage point from which to observe, or might it have led to omissions?

- Are the social relationships of the observer free from associations that might induce bias?

- Does the report appear to be self-serving? Does it benefit the experimenter to the extent that it should be questioned?

- Is the report internally consistent? Do the facts within the report support each other?

- Is the report externally consistent? Do the facts in the report agree with other independent reports?

As a result it is important to be explicit about data collection methods, the position of the researcher with respect to the subject matter, analysis processes, and codes. These details make it possible for other researchers to verify results.

In qualitative research it is acknowledged that the researcher's views, research context, and interpretations are an essential part of the qualitative research method as long as they are grounded in the collected data [3]. This does not, however, mean that qualitative evaluations are less trustworthy compared to quantitative research. Auerbach suggests using the concept of 'transferability' rather than 'generalizability' when thinking about the concepts of reliability and validity in qualitative research [3]. It is more important that the theoretical understanding we have gained can also be found in other research situations or systems and can be extended and developed further when applied to other scenarios. This stands in contrast to the concept of generalizability in quantitative research that wants to prove statistically that the results are universally applicable within the population under study.

Sometimes the point has been raised that if results do not generalize how can they be of use when designing software for general use. For example, qualitative methods might be used to obtain a rich description of a particular situation perhaps only observing the processes of two or three people. The results of a study like this may or may not generalize and the study itself provides no proof that they do. What we have is existence proof: that such processes are in use in at least two or three instances. Consider the worst case; that is that this rich description is an outlier that occurs only rarely. For design purposes, outliers are also important and sensitive design for outliers has been often shown to create better designs for all. For example, motion sensors to open doors may have been designed for wheelchairs but actually are useful features for all.

5.3.3 Analyzing Qualitative Data

Qualitative data may be analyzed using qualitative, quantitative, or a combination of both methods. Mixed methods research includes a qualitative phase and a quantitative phase in the overall research study in order to triangulate results from different methods, to complement results from one method with another, or to increase the breadth and range of inquiry by using different methods [28].

Many of the qualitative analysis methods can be grouped as types of thematic analysis, in which analysis starts from observations, then themes are sensed through review of the data, and finally coded [10]. Coding is the process of subdividing and labeling raw data, then reintegrating collected codes to form a theory [70]. Moving from the raw data into themes and a code set may proceed using one of three approaches: data-driven, motivated from previous research, or theory-driven, each with respectively decreasing levels of sensitivity to the data [10]. In the first style, data-driven, commonly called open coding [14]; themes and a code set are derived directly from the data and nothing else. If the analysis is motivated by previous research, the questions and perhaps codes from the earlier research can be applied to the new data to verify, extend or contrast the previous results. With theory-driven coding one may think using a given theory, such as grounded theory [13], or ethno-methodology [24], as a lens through which to view the data.

In either case the coded data may then be interpreted in more generalized terms. Qualitatively coded data may then be used with quantitative or statistical measures to try and distinguish themes or sampling groups.

5.4   Qualitative Summary

Qualitative studies can be a powerful methodology by which one can capture salient aspects of a problem that may provide useful design and evaluation criteria. Quantitative evaluation is naturally precision-oriented, but a shift from high precision to high fidelity may be made with the addition of qualitative evaluations. In particular, while qualitative evaluations can be used throughout the entire development life cycle in other research areas such as CSCW [41, 52, 64, 73], observational studies have been found to be especially useful for informing design. Yet these techniques are under-used and under-reported in the information visualization literature. Broader approaches to evaluation, different units of analysis and sensitivity to context are important when complex issues such as insight, discovery, confidence and collaboration need to be assessed. In more general terms, we would like to draw attention to qualitative research approaches which may help to address difficult types of evaluation questions. As noted by Isenberg el al. [36], a sign in Albert Einstein's office which read, *'Everything that can be counted does not necessarily count; everything that counts cannot necessarily be counted'* is particularly salient to this discussion in reminding us to include empirical research about important data that can not necessarily be counted.

6   Conclusions

In this paper we have made a two-pronged call: one for more evaluations in general and one for a broader appreciation of the variety of and importance of many different types of empirical methodologies. To achieve this, we as a research community need to both conduct more empirical research and to be more welcoming of this research in our publication venues. As noted in Section 4, even empirical laboratory experiments, as our most known type of empirical methodology, are often difficult to publish. One factor in this is that no empirical method is perfect. That is, there is always a trade-off between generalizability, precision, and realism. An inexperienced reviewer may recommend rejection based on the fact that one of these factors is not present, while realistically at least one will always be compromised. Empirical research is a slow, labour-intensive process in which understanding and insight can develop through time. That said, there are several important factors to consider when publishing empirical research. These include:

- That the empirical methodology was sensitively chosen. The methodology should be a good fit to the research question, the situation and the research goals.

- That the study was conducted with appropriate rigor. All methodologies have their own requirements for rigor and these should be followed. However, while trying to fit the rigor from one methodology onto another is not appropriate, developing hybrid methodologies that better fit a given research situation and benefit from two or more methodologies should be encouraged.

- That sufficient details are published so that the reader can fully understand the processes and if appropriate, reproduce them.

- That the claims should be made appropriately according to the strengths of the chosen methodology. For instance, if a given methodology does not generalize well, then generalizations should not be drawn from the results.

While there is growing recognition in our research community that evaluation information visualization is difficult [55, 57, 67], the recognition of this difficulty has not in itself provided immediate answers of how to approach this problem. Two positive recent trends of note are: one, that more evaluative papers in the form of usability studies have been published [25, 40, 47, 63, 80, 82], and two, that there are several papers that have made a call for more qualitative evaluations and complementary qualitative and quantitative approaches [18, 36, 48, 74].

This paper is intended merely as a pointer to a greater variety of empirical methodologies and encouragement towards their appreciation and even better their active use. There are many more such techniques and these types of techniques are being developed and improved continuously. There are good benefits to be had through active borrowing from ethnographic and sociological research methods, and applying them to our information visualization needs. In this paper we have argued for an increased awareness of empirical research. We have discussed the relationship of empirical research to information visualization and have made a call for a more sensitive application of this type of research [27]. In particular, we encourage thoughtful application of a greater variety of evaluative research methodologies in information visualization.

Acknowledgments. The ideas presented in this paper have evolved out of many discussions with many people. In particular this includes: Christopher Collins, Marian Dörk, Saul Greenberg, Carl Gutwin, Mark S. Hancock, Uta Hinrichs, Petra Isenberg, Stacey Scott, Amy Voida, and Torre Zuk.

References

1. Amar, R.A., Stasko, J.T.: Knowledge Precepts for Design and Evaluation of Information Visualizations. IEEE Transactions on Visualization and Computer Graphics 11(4), 432–442 (2005)
2. Andrews, K.: Evaluating Information Visualisations. In: Proceedings of the 2006 AVI Workshop on BEyond Time and Errors: Novel Evaluation Methods for Information Visualization, pp. 1–5 (2006)
3. Auerbach, C.: Qualitative Data: An Introduction to Coding and Analysis. University Press, New York (2003)

4. Baker, K., Greenberg, S., Gutwin, C.: Empirical Development of a Heuristic Evaluation Methodology for Shared Workspace Groupware. In: Proceedings of the ACM Conference on Computer Supported Cooperative Work, pp. 96–105. ACM Press, New York (2002)
5. Baldonado, M., Woodruff, A., Kuchinsky, A.: Guidelines for Using Multiple Views in Information Visualization. In: Proeedings of the Conference on Advanced Visual Interfaces (AVI), pp. 110–119. ACM Press, New York (2000)
6. Barzun, J., Graff, H.: The Modern Researcher, 3rd edn. Harcourt Brace Jovanvich, New York (1977)
7. BELIV 2006, accessed http://www.dis.uniroma1.it/~beliv06/ (February 4, 2008)
8. Bertin, J.: Semiology of Graphics (Translation: William J. Berg). University of Wisconsin Press (1983)
9. Bevan, N., Barnum, C., Cockton, G., Nielsen, J., Spool, J., Wixon, W.: The "Magic Number 5": Is It Enough for Web Testing? In: CHI Extended Abstracts, pp. 698–699. ACM Press, New York (2003)
10. Boyatzis, R.: Transforming Qualitative Information: Thematic Analysis and Code Development. Sage Publications, London (1998)
11. Brereton, M., McGarry, B.: An Observational Study of How Objects Support Engineering Design Thinking and Communication: Implications for the Design of Tangible Media. In: Proceedings of the ACM Conference on Human Factors in Computing Systems (CHI'00), pp. 217–224. ACM Press, New York (2000)
12. Chen, C., Czerwinski, M.: Introduction to the Special Issue on Empirical Evaluation of Information Visualizations. International Journal of Human-Computer Studies 53(5), 631–635 (2000)
13. Corbin, J., Strauss, A.: Basics of Qualitative Research: Techniques and Procedures for Developing Grounded Theory, 3rd edn. Sage Publications, Los Angeles (2008)
14. Creswell, J.: Qualitative Inquiry and Research Design: Choosing Among Five Traditions. Sage Publications, London (1998)
15. Dix, A., Finlay, J., Abowd, G., Beale, R.: Human Computer Interaction, 2nd edn. Prentice-Hall, Englewood Cliffs (1998)
16. Dumais, S., Cutrell, E., Chen, H.: Optimizing Search by Showing Results In Context. In: Proc. CHI'01, pp. 277–284. ACM Press, New York (2001)
17. Eberts, R.E.: User Interface Design. Prentice-Hall, Englewood Cliffs (1994)
18. Ellis, E., Dix, A.: An Explorative Analysis of User Evaluation Studies in Information Visualization. In: Proceedings of the Workshop on Beyond Time and Errors: Novel Evaluation Methods for Information Visualization, BELIV (2006)
19. Ericsson, K., Simon, H.: Verbal Reports as Data. Psychological Review 87(3), 215–251 (1980)
20. Ericsson, K., Simon, H.: Verbal Reports on Thinking. In: Faerch, C., Kasper, G. (eds.) Introspection in Second Language Research, pp. 24–54. Multilingual Matters, Clevedon, Avon (1987)
21. Ericsson, K., Simon, H.: Protocol Analysis: Verbal Reports as Data, 2nd edn. MIT Press, Boston (1993)
22. Fall, J., Fall, A.: SELES: A Spatially Explicit Landscape Event Simulator. In: Proceedings of GIS and Environmental Modeling, pp. 104–112. National Center for Geographic Information and Analysis (1996)
23. Forlines, C., Shen, C., Wigdor, D., Balakrishnan, R.: Exploring the effects of group size and display configuration on visual search. In: Computer Supported Cooperative Work 2006 Conference Proceedings, pp. 11–20 (2006)
24. Garfinkel, H.: Studies in Ethnomethodology. Polity Press, Cambridge (1967)

25. Gonzalez, V., Kobsa, A.: A Workplace Study of the Adoption of Information Visualization systems. In: Proceedings of the International Conference on Knowledge Management, pp. 92–102 (2003)
26. Gorard, S.: Combining Methods in Educational Research. McGraw-Hill, New York (2004)
27. Greenberg, S., Buxton, B.: Usability Evaluation Considered Harmful (Some of the Time). In: Proceedings of the SIGCHI Conference on Human Factors in Computing Systems (2008)
28. Greene, J., Caracelli, V., Graham, W.: Toward a Conceptual Framework for Mixed-Method Evaluation Design. Educational Evaluation and Policy Analysis 11(3), 255–274 (1989)
29. Gutwin, C., Greenberg, S.: The Mechanics of Collaboration: Developing Low Cost Usability Evaluation Methods for Shared Workspaces. In: Proceedings WETICE, pp. 98–103. IEEE Computer Society Press, Los Alamitos (2000)
30. Healey, C.G.: On the Use of Perceptual Cues and Data Mining for Effective Visualization of Scientific Datasets. In: Proceedings of Graphics Interface, pp. 177–184 (1998)
31. Heer, J., Viegas, F., Wattenberg, M.: Voyagers and Voyeurs: Supporting Asynchronous Collaborative Information Visualization. In: Proceedings of the Conference on Human Factors in Computing Systems (CHI'07), pp. 1029–1038. ACM Press, New York (2007)
32. Holtzblatt, K., Beyer, H.: Contextual Design: Defining Customer-Centered Systems. Morgan Kaufmann, San Francisco (1998)
33. Huck, S.W.: Reading Statistics and Research, 4th edn. Pearson Education Inc., Boston (2004)
34. Interrante, V.: Illustrating Surface Shape in Volume Data via Principal Direction-Driven 3D Line Integral Convolution. Computer Graphics, Annual Conference Series, pp. 109–116 (1997)
35. Isenberg, P., Carpendale, S.: Interactive Tree Comparison for Co-located Collaborative Information Visualization. IEEE Transactions on Visualization and Computer Graphics 12(5) (2007)
36. Isenberg, P., Tang, A., Carpendale, S.: An Exploratory Study of Visual Information Analysis. In: Proceedings of the Conference on Human Factors in Computing Systems (CHI'08), ACM Press, New York (to appear, 2008)
37. Kay, J., Reiger, H., Boyle, M., Francis, G.: An Ecosystem Approach for Sustainability: Addressing the Challenge of Complexity. Futures 31(7), 721–742 (1999)
38. Kim, S., Hagh-Shenas, H., Interrante, V.: Conveying Shape with Texture: Experimental Investigations of Texture's Effects on Shape Categorization Judgments. IEEE Transactions on Visualization and Computer Graphics 10(4), 471–483 (2004)
39. Kleffner, D.A., Ramachandran, V.S.: On the Perception of Shape from Shading. Perception and Psychophysics 52(1), 18–36 (1992)
40. Kobsa, A.: User Experiments with Tree Visualization Systems. In: Proceedings of the IEEE Symposium on Information Visualization, pp. 9–26 (2004)
41. Kruger, R., Carpendale, S., Scott, S.D., Greenberg, S.: Roles of Orientation in Tabletop Collaboration: Comprehension, Coordination and Communication. Journal of Computer Supported Collaborative Work 13(5–6), 501–537 (2004)
42. Lewin, C. (ed.): Research Methods in the Social Sciences. Sage Publications, London (2004)
43. Lewis, C., Rieman, J.: Task-Centered User Interface Design: A Practical Introduction (1993)
44. Likert, R.: A Technique for the Measurement of Attitudes. Archives of Psychology 140, 1–55 (1932)
45. Lofland, J., Lofland, L.: Analyzing Social Settings: A Guide to Qualitative Observation and Analysis. Wadsworth Publishing Company, CA, USA (1995)
46. Mankoff, J., Dey, A., Hsieh, G., Kientz, J., Lederer, S., Ames, A.: Heuristic Evaluation of Ambient Displays. In: Proceedings of CHI '03, pp. 169–176. ACM Press, New York (2003)
47. Mark, G., Kobsa, A., Gonzalez, V.: Do Four Eyes See Better Than Two? Collaborative Versus Individual Discovery in Data Visualization Systems. In: Proceedings of the IEEE Conference on Information Visualization (IV'02), July 2002, pp. 249–255. IEEE Press, Los Alamitos (2002)

48. Mazza, R., Berre, A.: Focus Group Methodology for Evaluating Information Visualization Techniques and Tools. In: Proceedings of the International Conference on Information Visualization IV (2007)
49. McCarthy, D.: Normal Science and Post-Normal Inquiry: A Context for Methodology (2004)
50. McGrath, J.: Methodology Matters: Doing Research in the Social and Behavioural Sciences. In: Readings in Human-Computer Interaction: Toward the Year 2000, Morgan Kaufmann, San Francisco (1995)
51. Moggridge, B.: Design Interactions. MIT Press, Cambridge (2006)
52. Morris, M.R., Ryall, K., Shen, C., Forlines, C., Vernier, F.: Beyond "Social Protocols": Multi-User Coordination Policies for Co-located Groupware. In: Proceedings of the ACM Conference on Computer-Supported Cooperative Work (CSCW, Chicago, IL, USA), CHI Letters, November 6-10, 2004, pp. 262–265. ACM Press, New York (2004)
53. Morse, E., Lewis, M., Olsen, K.: Evaluating Visualizations: Using a Taxonomic Guide. Int. J. Human-Computer Studies 53, 637–662 (2000)
54. Nielsen, J., Mack, R.: Usability Inspection Methods. John Wiley & Sons, Chichester (1994)
55. North, C.: Toward Measuring Visualization Insight. IEEE Computer Graphics and Applications 26(3), 6–9 (2006)
56. Patton, M.Q.: Qualitative Research and Evaluation Methods, 3rd edn. Sage Publications, London (2001)
57. Plaisant, C.: The Challenge of Information Visualization Evaluation. In: Proceedings of the Working Conference on Advanced Visual Interfaces, pp. 109–116 (2004)
58. Purchase, H.C., Hoggan, E., Görg, C.: How Important Is the "Mental Map"? – An Empirical Investigation of a Dynamic Graph Layout Algorithm. In: Kaufmann, M., Wagner, D. (eds.) GD 2006. LNCS, vol. 4372, pp. 184–195. Springer, Heidelberg (2007)
59. Purchase, H.C.: Effective Information Visualisation: A Study of Graph Drawing Aesthetics and Algorithms. Interacting with Computers 13(2), 477–506 (2000)
60. Purchase, H.C.: Performance of Layout Algorithms: Comprehension, Not Computation. Journal of Visual Languages and Computing 9, 647–657 (1998)
61. Brandenburg, F.J. (ed.): GD 1995. LNCS, vol. 1027. Springer, Heidelberg (1996)
62. Reilly, D., Inkpen, K.: White Rooms and Morphing Don't Mix: Setting and the Evaluation of Visualization Techniques. In: Proceedings of the SIGCHI Conference on Human Factors in Computing Systems, pp. 111–120 (2007)
63. Saraiya, P., North, C., Duca, K.: An Insight-Based Methodology for Evaluating Bioinformatics Visualizations. IEEE Transactions on Visualization and Computer Graphics 11(4), 443–456 (2005)
64. Scott, S.D., Carpendale, S., Inkpen, K.: Territoriality in Collaborative Tabletop Workspaces. In: Proceedings of the ACM Conference on Computer-Supported Cooperative Work (CSCW, Chicago, IL, USA), CHI Letters, November 6-10, 2004, pp. 294–303. ACM Press, New York (2004)
65. Seidman, I.: Interviewing as Qualitative Research: A Guide for Researchers in Education and the Social Sciences. Teachers' College Press, New York (1998)
66. Shneiderman, B.: The Eyes Have It: A Task by Data Type Taxonomy for Information Visualizations. In: Proceedings of the IEEE Symposium on Visual Languages, pp. 336–343. IEEE Computer Society Press, Los Alamitos (1996)
67. Shneiderman, B., Plaisant, C.: Strategies for Evaluating Information Visualization Tools: Multi-Dimensional In-Depth Long-Term Case Studies. In: Proceedings of the Workshop on BEyond Time and Errors: Novel Evaluation Methods for Information Visualization, BELIV (2006)
68. Spence, R.: Information Visualization, 2nd edn. Addison-Wesley, Reading (2007)
69. Spool, J., Schroeder, W.: Testing Web Sites: Five Users is Nowhere Near Enough. In: CHI '01 Extended Abstracts, pp. 285–286. ACM Press, New York (2001)

70. Strauss, A.L., Corbin, J.: Basics of Qualitative Research: Techniques and Procedures for Developing Grounded Theory. Sage Publications, London (1998)
71. Tang, A., Tory, M., Po, B., Neumann, P., Carpendale, S.: Collaborative Coupling over Tabletop Displays. In: Proceedings of the Conference on Human Factors in Computing Systems (CHI'06), pp. 1181–1290. ACM Press, New York (2006)
72. Tang, A., Carpendale, S.: An observational study on information flow during nurses' shift change. In: Proc. of the ACM Conf. on Human Factors in Computing Systems (CHI), pp. 219–228. ACM Press, New York (2007)
73. Tang, J.C.: Findings from observational studies of collaborative work. International Journal of Man-Machine Studies 34(2), 143–160 (1991)
74. Tory, M., Möller, T.: Evaluating Visualizations: Do Expert Reviews Work. IEEE Computer Graphics and Applications 25(5), 8–11 (2005)
75. Tufte, E.: The Visual Display of Quantitative Information. Graphics Press, Cheshire (1986)
76. Tufte, E.: Envisioning Information. Graphics Press, Cheshire (1990)
77. Tufte, E.: Visual Explanations. Images and Quantities, Evidence and Narrative. Graphics Press, Cheshire (1997)
78. Viegas, F.B., Wattenberg, M., van Ham, F., Kriss, J., McKeon, M.: Many Eyes: A Site for Visualization at Internet Scale. IEEE Transactions on Visualization and Computer Graphics (Proceedings Visualization / Information Visualization 2007) 12(5), 1121–1128 (2007)
79. Ware, C.: Information Visualization: Perception for Design, 2nd edn. Morgan Kaufmann, San Francisco (2004)
80. Wigdor, D., Shen, C., Forlines, C., Balakrishnan, R.: Perception of Elementary Graphical Elements in Tabletop and Multi-surface Environments. In: Proceedings of the Conference on Human Factors in Computing Systems (CHI'07), pp. 473–482. ACM Press, New York (2007)
81. Willett, W., Heer, J., Agrawala, M.: Scented Widgets: Improving Navigation Cues with Embedded Visualizations. In: INFOVIS 2007. IEEE Symposium on Information Visualization (2007)
82. Yost, B., North, C.: The Perceptual Scalability of Visualization. IEEE Transactions on Visualization and Computer Graphics 12(5), 837–844 (2006)
83. Zuk, T.: Uncertainty Visualizations. PhD thesis. Department of Compute Science, University of Calgary (2007)
84. Zuk, T., Carpendale, S.: Theoretical Analysis of Uncertainty Visualizations. In: Proceedings of SPIE Conference Electronic Imaging, Vol. 6060: Visualization and Data Analysis (2006)
85. Zuk, T., Schlesier, L., Neumann, P., Hancock, M.S., Carpendale, S.: Heuristics for Information Visualization Evaluation. In: Proceedings of the Workshop BEyond Time and Errors: Novel Evaluation Methods for Information Visualization (BELIV 2006), held in conjunction with the Working Conference on Advanced Visual Interfaces (AVI 2006), ACM Press, New York (2006)

Theoretical Foundations of Information Visualization

Helen C. Purchase[1], Natalia Andrienko[2], T.J. Jankun-Kelly[3], and Matthew Ward[4]

[1] Department of Computing Science, University of Glasgow,
17 Lilybank Gardens, G12 8QQ, UK,
hcp@dcs.gla.ac.uk
[2] Fraunhofer Institut Intelligent Analysis & Information Systems (FhG IAIS),
Schloss Birlinghoven, D-53754 Sankt-Augustin, Germany,
natalia.andrienko@iais.fraunhofer.de
[3] Department of Computer Science and Engineering,
Bagley College of Engineering, Mississippi State University,
Mississippi State, MS 39762, USA,
tjk@acm.org
[4] Computer Science Department, Worcester Polytechnic Institute,
100 Institute Road, Worcester, MA 01609-2280, USA,
matt@cs.wpi.edu

Abstract. The field of Information Visualization, being related to many other diverse disciplines (for example, engineering, graphics, statistical modeling) suffers from not being based on a clear underlying theory. The absence of a framework for Information Visualization makes the significance of achievements in this area difficult to describe, validate and defend. Drawing on theories within associated disciplines, three different approaches to theoretical foundations of Information Visualization are presented here: data-centric predictive theory, information theory, and scientific modeling. Definitions from linguistic theory are used to provide an over-arching framework for these three approaches.

1 Introduction

Information Visualization suffers from not being based on a clearly defined underlying theory, making the tools we produce difficult to validate and defend, and meaning that the worth of a new visualization method cannot be predicted in advance of implementation. There is much unease in the community as to the lack of theoretical basis for the many impressive and useful tools that are designed, implemented and evaluated by Information Visualization researchers.

The purpose of a theory is to provide a framework within which to explain phenomena. This framework can then be used to both evaluate and predict events, in this case, users' insight or understanding of visualization, and their use of it. An Information Visualization theory would enable us to evaluate visualizations with reference to an established and agreed framework, and to predict the effect of a novel visualization method.

This is not to say that a single theory would be able to encapsulate the whole of the Information Visualization field; it may be that multiple theories at different levels are needed. We already make use of many existing cognitive and perceptual theories, as well as established statistical methods. It might be that the complexity of Informa-

A. Kerren et al. (Eds.): Information Visualization, LNCS 4950, pp. 46–64, 2008.
© Springer-Verlag Berlin Heidelberg 2008

tion Visualization as relating to engineering, cognition, design and science requires the use of several theories each taking a different perspective.

As a starting point, we liken the understanding of visualization to the understanding of ideas expressed in language. We draw on two perspectives in linguistic theory: language as representation and language as process.

In considering the representation of language, lexical tokens are syntactically ordered to produce a semantic concept which the reader understands with reference to a learned code. The concept is understood within a context and the reader responds pragmatically.

We can take this approach a step further and consider the semiotic theory of Saussure [1] wherein a sign is a relation between a perceptible token (signifier, referrer) and a concept (signified, referent) – giving us another useful term: the referent of the token, i.e. what it actually means in a given context. For example, a pair of numbers (the referrer) may mean a geographical location in one context (one possible referent), yet may mean a student's examination and coursework marks in another (a different referent).

Similarly, we can extend our consideration of pragmatics to include stylistics: the style in which language is written. The same pragmatic response may be stimulated by a different set of tokens (or the same set arranged with a different syntax) – this may produce a different emotional response.

This Saussurian view of language as a static representation of meaning can be contrasted with the view of Bhaktin [2] who considers language to be a dynamic process whereby a text is interacted with and manipulated, and its meaning constructed dynamically. This active and engaged understanding, Bhaktin says, creates new meanings: "… establishes a series of complex interrelationships, consonances and dissonances … [and] various different points of view, conceptual horizons … come to interact with one another" [2].

While Bhaktin's theory of the dynamic interpretation and negotiation of linguistic texts was primarily based around the social context of language interpretation and the construction of ideologies within cultures and institutions, it is a useful complement to the "language as representation" perspective presented above. We can use this alternate view of "language as a process" in our framework for Information Visualization: a model of data embodied in a visualization must be explored, manipulated and adapted within a investigatory process resulting in enhanced understanding of its meaning. When there are two processing agents in a human-computer interaction context (the human and the computer), either or both can perform this processing.

Thus, we can relate discussion of theoretical approaches to Information Visualization to the concepts of

- Interpretation of a visualization through its external physical form (referents, and lexical, syntactic, semantic, pragmatic and stylistic structures), an activity typically performed by a reader;
- Exploration and manipulation of the external representation by the reader so as to discover more about the underlying model, typically done through interaction facilities provided by a visualization tool; and
- Exploration and manipulation of the internal data model by the system in order to discover interrelationships, trends and patterns, so as to enable them to be represented appropriately.

The three sections that follow each take a different approach to suggesting a theory for Information Visualization. While they were not originally developed with the above linguistic model in mind, each can be related in some way to this framework.

Natalia Andrienko takes a data-centric view, focusing on the dataset itself, and the tokens that describe it. She considers how the characteristics of the dataset and the requirements of the visualization for a task may be matched to determine patterns, thus predicting the most appropriate visualization tool for the given task. Thus, this section describes the exploration of the data model so as to identify the best syntax to use for given tokens (taking into account their referents and the desired semantics). She highlights the usefulness of systems which can explore the data model, predict the patterns in datasets, and facilitate the perception of these patterns.

Matthew Ward's starting point is communication theory, and this section is clearly focused on information content – the meaning of the visualization and maintaining the flow of information through all stages of the visualization pipeline. He discusses how we may assess our progress in designing and enhancing visualizations through considering measurements of information transfer, content or loss, thus providing a useful theoretical means for validating visualizations. In this case, there is no internal exploration of the data, but it is the validity of the data after transfer from internal model to external representation that is considered important.

T.J. Jankun-Kelly introduces two useful models for a scientific approach to visualization, both of which are in their infancy. The visual exploration model describes and captures the dynamic process of user exploration and manipulation of visualization in order to affect its redesign, thus using the pragmatic response of the user to determine a new syntactical arrangement. The second model, visual transformation design, uses transformation functions applied to the data model to provide design guidance based on visualization parameters, thus performing an initial exploration of the data model to suggest syntax to enhance the pragmatic response of the user.

The paper concludes with a summary, and suggestions for future research.

2   Predictive Data-Centered Theory

Among other theories, Information Visualization requires a theory that could serve as a basis for instructing Information Visualization users how to select the right tools for their data and do data exploration and analysis with the use of these tools. The same theory could also help tool designers in finding right solutions. The following argumentation is meant to clarify what kind of theory this could be.

Most Information Visualization researchers agree that the primary purpose for using Information Visualization tools is to explore data in order to gain understanding of the data and the phenomena behind. Gaining understanding may be thought of as constructing a concept, or mental model, of the data or phenomenon. A model, in turn, can be considered as a parsimonious representation capturing essential features of the data rather than listing all individual data items; this means that a model necessarily involves abstraction. For example, from observing morning temperatures over several days, a person may build a concept of the increase or decrease of the temperature.

Such an abstraction is based on a holistic grasp of characteristic features embracing multiple data items. We shall use the term "pattern" to refer to such features. Increase and decrease are examples of patterns. A model may be a synthesis of several patterns each representing some part or aspect of the data. Thus, when the observation of the morning temperatures is performed over a sufficiently long period, the model will probably incorporate the patterns of both increase and decrease of the temperature. Furthermore, patterns may also be composed of sub-patterns. For instance, the behavior of the temperature may be conceptualized as a repeated "wave" where increase is followed by decrease. Here, increase and decrease are basic, or atomic, patterns, the "wave" is a composite pattern including the increase and decrease patterns, and the repetition of the "wave" is a pattern of a yet higher level, which incorporates the "wave" pattern.

The main role of Information Visualization tools can be understood as helping the user to perceive patterns that could be used for building an appropriate model. This means, in particular, that a tool should facilitate the perception of (sub)sets of data items as units. For an appropriate support of the detection of patterns, a tool designer should know in advance what *types* of patterns need to be perceived (or otherwise detected) with the use of the tool. Then, after the tool is ready, it will be easy to explain to the users the purpose of the tool and instruct them how to detect the types of patterns the tool is oriented to.

The types of patterns that may be meaningful for the user depend on the structure and properties of the data under analysis. Thus, in the analysis of a temporal series of numeric measurements (such as temperatures) it makes sense to look for such basic patterns as increase, decrease, stability, fluctuation, peak, and low point. However, when numeric measurements refer to a discrete unordered set as, for example, melting temperatures of various substances, the possible types of patterns may be groupings of elements with close values of the measurements and frequency-related patterns: prevalence of certain values or value intervals, frequent values or exceptional values (outliers).

To support the designers and users of Information Visualization tools in the way described above, there is a need for a theory that could enable the possibility to predict, for a given dataset or a given class of datasets, what *types* of patterns may be found there. We specially emphasize the term *types* to exclude the possible impression of attempting to predict (and on this basis automatically detect) all specific patterns hidden in specific data. Thus, a prediction that a dataset may contain groups (clusters) of objects with similar characteristics does not define what specific clusters are there. However, it orients tool designers, who will know that the tool must help the users to detect clusters, and users, who will know that they need a tool facilitating the detection of clusters. Then, if each Information Visualization tool and technique is supplied with an appropriate signature (i.e. what kind of data it is suitable for and what types of patterns it is oriented to), the user will be able to choose the right tool.

The theory we are advocating in this section can be called data-centered predictive theory. The theory needs to include

1. an appropriate generic framework for the characterization of various data types and structures;
2. a general typology of patterns;
3. a mechanism for deriving possible pattern types from data characterizations.

Here, we present some preliminary ideas concerning these components of the theory.

2.1   Data Characterization Framework

Data may be viewed abstractly as a set of records with a common structure, each record being a sequence of elements (such as numbers or strings) which either reflect the results of some observations or measurements or specify the context in which the observations or measurements were obtained. The context may include, for example, the place and the time of observation or measurement, and the object or group of objects observed. The elements that a data record consists of are called *values*.

All records of a dataset are assumed to have a common structure, with each position having its specific meaning, which is common to all values appearing in it. These positions may be named to distinguish between them. The positions are usually called *components* of the data.

Definition: *Characteristic component*, or *attribute*, is a data component corresponding to a measured or observed property of the phenomenon reflected in the data. *Characteristic* is a value of a single attribute or a combination of values of several dataset attributes.

Definition: *Referential component*, or *referrer*, is a data component reflecting an aspect of the context in which the observations or measurements were made. *Reference* is the value of a single referrer or the combination of values of several referrers that fully specifies the context of some observation(s) or measurement(s).

Definition: *Reference set* of a dataset is the set of all references occurring in this dataset.

Definition: *Characteristic set* of a dataset is the set of all possible characteristics, (i.e. combinations of values of the dataset attributes).

Definition: *Multidimensional dataset* is a dataset having two or more referrers. Depending on the number of referrers, a dataset may be called one-dimensional, two-dimensional, three-dimensional, and so on.

For example, the geographical location and the time are referrers for measurements of properties of the climate such as air temperature or wind direction, which are attributes. Each combination of location and time is a reference, and the corresponding combination of air temperature and wind direction is a characteristic. This is a two-dimensional dataset as it has two referrers; the attributes are not counted as dimensions. Referrers are *independent* components and attributes are *dependent* since the values of attributes depend on the context in which they are observed. In data analysis, it is possible to deal with selected attributes independently from the others; however, all referrers present in a dataset need to be handled simultaneously.

Data may be viewed formally as a function, in the mathematical sense, with the referrers being independent variables and the attributes being dependent variables. The function defines the correspondence between the references and the characteristics where for each combination of values of the referential components there is at most one combination of values of the attributes.

The structure of a dataset is characterized by specifying which components it includes, which of them are referrers, and which ones are attributes. Additionally to this, it is necessary to specify the properties of the components. The relevant properties are:

- whether distances exist between the elements. Any continuous set such as time, space, and values of temperature has distances, but there may be distances also in discrete sets such as a set of integer values denoting numbers of some items. The discrete set of substances has no distances.
- whether and how the elements are ordered. Thus, time moments are linearly ordered and may also be cyclically ordered, depending on the time span of observations.

It should be noted that a set consisting of combinations of values of several components does not inherit the properties of the individual components. Thus, a set of combinations of values of melting temperature and atomic weight is only partly ordered although the value sets of the original attributes are fully ordered. This data characterization framework is presented in more detail in [3].

2.2   Patterns

Definition: *Pattern* is an artifact that represents some arrangement of characteristics corresponding to a (sub)set of references in a holistic way, i.e. abstracted from the individual references and characteristics.

This is a more generic definition than is given in data mining: "a pattern is an expression E in some language L describing facts in a subset F_E of a set of facts F [i.e. a dataset, in our terms] so that E is simpler than the enumeration of all facts in F_E" [4]. In our definition, we mean any kind of representation, for example, graphical or mental.

We posit that all existing and imaginable patterns may be considered as instantiations of certain archetypes (or, simply, types). It is quite reasonable to assume that such archetypes may exist in the mind of a data analyst and drive the process of visual data analysis, which is commonly believed to be based on pattern recognition: the analyst looks for constructs that can be regarded as instances of the existing archetypes.

A pattern-instance may be characterized by referring to its type and specifying its individual properties, in particular, the reference (sub)set on which the pattern is based. Properties may be type-specific (for example, amount and rate of increase).

2.3   Pattern-by-Data Typology

The following table defines the basic types of patterns in relation to the characteristics of data for which such pattern types are relevant. We cannot guarantee at the present moment that this typology is complete; further work is obviously needed. Note that neither the columns nor the rows of the table are mutually exclusive. Thus, when the characteristic set is ordered and has distances, the pattern types from all columns are relevant. Similarly, when the reference set is linearly and cyclically ordered, the patterns from all rows are possible.

These basic pattern types may be included in composite patterns. The types of composite patterns depend on the properties of the reference set:

Characteristic set / Reference set	Any	Ordered	Has distances
Any	Even frequencies of the values, prevailing values, rare values	Tendency toward high, low or medium values	Groups (clusters) of references with close characteristics
Linearly ordered	Constancy, change, specific value order	Increase, decrease, peak, low point	Gradual change, sharp change
Cyclically ordered	Frequency of value appearing in certain positions of the cycle	Cyclical increase and decrease	Gradual or sharp changes within the cycle and between cycles
Has distances	Homogeneity or heterogeneity, large or small regions of congruency	Flatness, elevation, depression, peak, depth, plateau, valley	Smoothness (small differences between characteristics of neighboring references), abruptness (big differences)

1. For any kind of reference set: repeated pattern, frequent pattern, infrequent pattern, prevailing pattern;
2. For a linearly ordered reference set: specific sequence of patterns, alternation;
3. For a cyclically ordered reference set: cyclically repeated pattern;
4. For a reference set with distances: constant distance between repetitions of a pattern, patterns occurring close to each other.

Any composite pattern may, in turn, be included in a bigger composite pattern, for example, a frequently repeated pattern where increase is followed by decrease.

2.4   Directions of Further Work

What is presented in this section is only an initial sketch of the data-centered predictive theory. Further work is required to ensure the comprehensiveness of the pattern typology. Particular attention needs to be paid to multi-dimensional data. It is also necessary to define pattern types used to represent relationships between attributes or between phenomena (represented by several datasets differing in structure) such as correlation (co-occurrence) or influence.

Then, it is necessary to evaluate Information Visualization techniques according to the types of data they are suited for and the types of patterns they help to elicit. This can form an appropriate basis for instructive books and courses for users of Information Visualization tools.

3   Information Theory

3.1   Visual Communication

Information visualization can be viewed as a communication channel from a dataset to the cognitive processing center of the human observer. This suggests that it might be possible to employ concepts from the theories of data communication as a mechanism for evaluating and improving the effectiveness of information visualization techniques. While there are several early papers that tried to establish linkages between HCI in general to information theory [5], it might be time to revisit this concept in light of all the progress that has been made in information visualization in the past two decades.

We must start with defining information, as it is the core of information visualization.

Schneider defined information as "always a measure of the decrease of uncertainty at a receiver" [6] while Cherry stated "Information can only be received where there is doubt; and doubt implies the existence of alternatives where choice, selection, or discrimination is called for" [7]. Measuring information is a topic found in many fields, including science, engineering, mathematics, psychology, and linguistics. Information theory, which primarily evolved out of the study of hardware communication channels, defines entropy as the loss of information during transmission; it is also referred to as a measure of disorder or randomness. Another important term is bandwidth, which is a measure of the amount of information that can be delivered over a communication channel in a given period of time. We will attempt to analyze information visualization using this terminology.

Information can be categorized in a number of ways. MacKay [8] identifies three types of information content:

- Selective information-content: This is information that helps the receiver make a selection from a set of possibilities, or narrows the range of possibilities. An example might be a symptom that helps make a diagnosis.
- Descriptive information-content: This type of information helps the user build a mental model. Two types of descriptive information content have been identified:
  - Metrical: this type of observation increases the reliability of a pattern, e.g., a new member of an existing cluster (sometimes termed a metron).
  - Structural: this type of observation adds new features or patterns to a model/representation, e.g., a new cluster (termed a logon).
- Semantic information-content: This type of information is not covered in classical information theory. It lies between the physical signals of communications (called the syntactic) and the users and how they respond to the signals (called the pragmatics). The pragmatics are the domain of psychology, both perceptual and cognitive.

While the first two classes of information content lend themselves well to measurement, it is much harder to determine measures of semantic content, as this in general is very specific to individuals.

3.2   Measuring the Amount of Information

There have been many efforts to date to quantify the amount of information in a communication stream. If we think of plain text, there are numerous quantifiable features, including:

- The total number of words per minute
- The occurrence of specific words
- The frequency of occurrence for each word
- The occurrence of word pairs, triples, phrases, and sentences.

There are problems, however, with such simplistic, syntax-only measurement. Words can have variable significance; some are unnecessary or redundant. Many words can encode the same concept. In fact, reading text or hearing speech may have no affect on one's uncertainty regarding the subject of the text, e.g., you may already have known it, or you don't understand the meaning of the words or their implied concepts. This implies that the measurement of information content or volume can be specific to the individual receiver and, as we'll see later, the task that is being performed based on the communication.

Can we perform similar analysis on a dataset? Consider a table of numeric values. Features of potential interest in the dataset include:

- The count of number of entries or dimensions
- The values
- Clusters and their attributes (number, size, relations, ...)
- Trends and their attributes (size, rate of change, ...)
- Outliers and their attributes (number, degree of outlierness, relation to dense regions, ...)
- Associations, correlations and any features between records, dimensions, or individual values.

In fact, we can observe that a featureless dataset is not differentiable from random noise: all values are equally likely. Features and relations can also vary in their magnitude, certainty, complexity, and importance. Clusters may differ in size; outliers may vary in their distance to the main body of data; features may be comprised of many sub-features; in many cases, a feature that is significant to one observer may be considered noise by another. Recently, researchers have proposed measuring and counting insights [9], which are new knowledge gained during visual analysis. These insights are generally specific to a particular task, some of which include [10]:

- Identify data characteristics
- Locate boundaries, critical points, other features
- Distinguish regions of different characteristics
- Categorize or classify
- Rank based on some order
- Compare to find similarities and differences
- Associate into relations
- Correlate by classifying relations.

For each of these tasks, we might have different accuracy requirements as well, which can influence the resolution at which feature extraction is accomplished during communication. Thus, for example, the tasks of detecting, classifying, and measuring a particular phenomenon each have their own accuracy demands. The tasks to be performed also have an implication on the types of information that the visualization must be able to convey; categorization and ranking imply that the visualization must have high selective information content, while identifying characteristics and boundaries are part of building a mental model and thus require good descriptive information content.

Returning to our dataset and the simplistic features and relations that are contained in it, we can try to quantify the volume of information and then measure how much of this volume a visualization technique is capable of effectively conveying. If we assume a table of scalar values (M records, N dimensions or variables), the number of individual values to be communicated is M*N, and the maximum resolution required is the number of significant digits. Often, however, the available visual resolution is far less than that of the data. We can then count all the pairwise relations between records, or dimensions, or even values. For records, this would be $M*(M-1)/2$, and similar for dimensions and values. Then there are relations that are 3-way, 4-way, or even among an arbitrary number of elements, e.g., in clustering tasks. Clearly, there are too many possibilities to consider them all, so perhaps we need a different tactic.

3.3   Measuring Information Loss

Perhaps it is easier to measure loss of information (entropy) during the visualization process than the total information content of a dataset. There are several techniques in common use for data transformation for visualization that provide an implicit measure of information loss. For example, multidimensional scaling, a process commonly used for dimensionality reduction, provides a measure of stress, which is the difference between the distances between points in the original dimensioned space and the corresponding distances in the reduced dimension space. Similarly, when using principal component analysis for performing this reduction, the loss can be measured from the dropped components. Cui et al. [11] developed measures of representativeness when using processes such as sampling and clustering to reduce the number of data records in the visualization. These measures, based on nearest neighbor computations, histogram comparisons, and statistical properties, give the analysts control over what was termed abstraction quality, so they are aware of the trade-offs between speed of rendering, display clutter, and information loss. They, however, did not consider the perceptual issues, which are very dependent on the particular visual encoding used.

Distortion techniques such as lens effects and occlusion reduction also provide the analyst with trade-offs between accuracy and visual clarity. Each results in a transformation (typically of an object's position on the screen) that is meant to improve the local interpretability at the cost of accuracy of global relations. It would be interesting to see measures of these competing processes to gauge the overall implications.

Another transformation that can impact on the information being communicated in a visualization is the ordering of records and dimensions. Ordering can reveal trends, associations, and other types of relations, and is useful for many tasks. There are many possible orderings of a table of M records and N dimensions. The key is to determine which are the most useful. An ordering can convey many pairwise relations. If there are M records, an ordering can communicate M-1 of the $M*(M-1)/2$ possible pairwise relations. Many researchers have studied ways of selecting a useful ordering, including Bertin's reorderable matrix [12], Seo and Shneiderman's rank-by-feature techniques [13], and Peng et al.'s reordering for visual clutter reduction [14]. In all cases, a user should be able to prune orderings to emphasize those that show trends, groupings, or other discernable patterns. Thus far most research has focused on simple 1-D orderings, but higher level orderings and structures (e.g., hierarchies) have also been studied.

3.4   Hardware and Perceptual Limitations

This discussion of information content would not be complete without also considering the limitations imposed by the visual communication channel (i.e., the display) and receiver (the human visual perception system). Regarding the channel and its capacity, modern displays are limited to somewhere on the order of one to nine million pixels, although tiled displays can increase this substantially. The color palette generally has a size of 2^{24} possible values, although the limitations of human color perception take a big chunk out of this. Finally, the refresh rate of the system, typically between 20 and 30 frames per second, limits how fast the values on the screen change, though again the human limitations of change detection mean that much of this capability is moot.

Regarding these human limitations, from the study of human physiology we know that there are approximately 800k fibers in the optic nerve. We can perceive 8-9 levels of intensity graduation, and require a 0.1 second accumulation period to register a visual stimulus. In addition, we have a limited viewable area at any particular time, and a variable density of receptors (much less dense in the peripheral vision). Studies have shown we have a limited ability to distinguish and measure size, position, and color, and the duration of exposure affects our capacity. Finally, it has been shown that our abilities are also related to the task at hand; we are much better at relative judgment tasks than absolute judgment ones.

3.5   Measuring Information Content on Visualizations

We now look at methods that have been used in the past for measuring the information content in a data or information visualization. For completeness sake, some of these are quite trivial. For example, simply counting the number of data values shown is a valid measure. The issue in this case would be how to deal with partial occlusion. In some cases this would be acceptable if sufficient information remains visible to make identification or recognition possible. Tufte [15] suggested the data-ink ratio as an indicator of information content, though tick marks, labels, and axes are often essential for appropriate identification. Many researchers have used counts of the number of features or patterns found in a particular amount of time. Ward and Theroux [16] counted the number of clusters and outliers found by users in different visualizations, while Suraiya et al. [9] counted insights discovered. In each case, a ground truth is needed to verify that what was found was really present. Similar experiments have been used to measure classification, measurement, and recall accuracy.

There are many other issues when attempting to measure information in a visualization. As mentioned earlier, distortion and other transformations can improve the readability of a visualization, but introduce errors in the data themselves. Data may have uncertainty attributes associated with them, which can interfere with the measurement. On the other hand, there are numerous examples of improving information content by using novel layout, shape, and color strategies or augmenting the visualization with links, boundaries, and even white space. The amount of information contained may also be enhanced using redundant mappings, which improves the chances of successful reception by the viewer. Finally, the use of animation to communicate information in an incremental fashion can be quite effective; it is lossy communication, as viewers quickly forget some of what they have seen, but the ability to replay the animation can replace some of this lost information.

3.6   Case Study: Parallel Coordinates

As an example of this information measurement activity, let us consider parallel coordinates, a popular multivariate visualization technique. The first question is how well does this technique present the values of a dataset? For individual values, this method has very high resolution, given most of the screen height can be used to convey the value. This implies the technique possesses high selective information content, at least on individual dimensions, as separation into sub-ranges is facilitated by the amount of screen space allocated to convey values. However, the loss due to occlusion can be high, especially for nominal variables. This loss is somewhat mitigated by the varying slopes of lines ending/starting at axes, which allows some degree of differentiation. Relationships among data along an axis can be emphasized via embedded histograms, as found in some implementations of parallel coordinates. In terms of relationships between dimensions, this method is limited to showing pairwise relations, with N-1 out of the N*(N-1)/2 possible relations shown. Automated dimension ordering can help reveal interesting relationships, as seen in [14]. Relationships between records are problematic due to the ambiguous continuity of records that intersect on one or more axes. Coloring the lines based on a record ID can help with modest sized datasets. We can also use animation along an axis or based on an order to expose some inter-record relations. Intersections and near-parallel edges can reveal partial structures (between dimension pairs), and techniques such as Hierarchical Parallel Coordinates [17] can show grouping relations, though there is loss of individual data values. Each of these methods enhances the descriptive information content of the visualization, thus helping analysts form mental models of the data.

Through analyzing these augmentations of parallel coordinates we see that many recent innovations to parallel coordinates target different types of information loss resulting from this means of mapping data. For example, we see efforts to preserve and emphasize outliers in a paper by Novotny and Hauser [18]. However, many other issues still exist; there are still many data features that cannot be readily perceived and tasks that are difficult to perform using parallel coordinates.

3.7   Conclusions

To summarize, we feel there are many aspects of information visualization research that can find analogies in the concepts of information theory - it is all about communication. Perhaps finding such a formal structure on which to ground our efforts can potentially reduce the amount of ad hocness in the field. The key is to define measures of information transfer, content, or loss at all stages of the pipeline as a means of assessing our progress in the development of new visualization techniques and enhancement of existing ones.

4   Formal Models for a Science of Information Visualization

Information visualization utilizes computer graphics and interaction to assist humans in solving problems. As such, it incorporates elements of a constructive, formal science (the algorithmic and development aspects) with aspects of an empirical science (for measuring effectiveness and validity). Surrounding both are engineering efforts to improve the overall system. This section discusses the relationship between these three parts of the visualization discipline, and suggests that a deeper exploration of formal, scientific models is needed for a strengthening of the field.

4.1   The Need for Models

Traditionally, visualization has focused on the engineering aspects while importing "scientific" elements as needed. However, even this borrowing has not been sufficiently utilized. To illustrate this, consider archetypical topics from an information visualization course syllabus:

- *Exploration:* The process of visual exploration in a larger context
- *Perception:* Fundamental mechanisms for human visual perception
- *Visual Cognition*: How perception translates to thought and action
- *Color:* Aspects of color for visualization
- *Techniques:* Specific visualization metaphors, including interaction
- *Evaluation:* Measuring the effectiveness of a visualization design

A survey of visualization education programs has found that most such programs focus on visualization techniques (the engineering core) in detriment to the foundational aspects (the scientific core) [19]. As a further example, consider that rainbow colormaps are still entrenched in visualization research [20] while ample scientific evidence demonstrates their muddling effect [20,21,22]. The ease of utility for providing rainbow colormaps does not outweigh the costs in terms of a user's time - the primary currency of users [23].

Examining the previous list, it is apparent that only the Techniques topic deals extensively with the engineering aspects of visualization design. While these efforts are vital to providing actual tools to users, the other elements are needed to provide a solid foundation to guide those efforts. For example, perceptual literature is grounded in empirical results with a strong scientific pedigree. A key aspect of these results is the *formal models* which are generated to explain the results. Such models are both *descriptive* - they encapsulate the factors of the empirical experiment and describe a mechanism for their operation - and *predictive* - they generalize the description to a larger context by predicting future behavior. The predictive nature of the models facilitates visualization design: it is the predictive nature of color perception models that explains the limitations of rainbow colormaps [22].

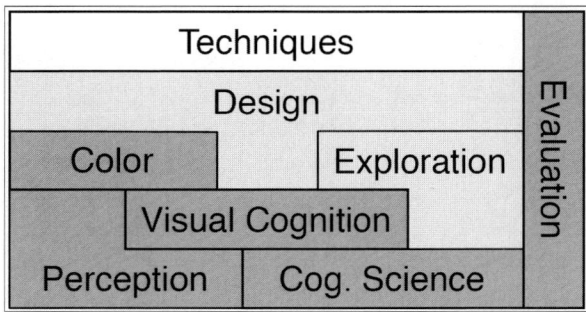

Fig. 1. Topics in a Formalized Information Visualization Course. Dark grey topics are based upon formal foundations in other disciplines; light grey topics are yet-to-be-developed visualization-specific formal foundations.

4.2   A Move to Visualization Formalisms: The Two Models

There have been several recent calls for an establishment of a "theory" and "science" behind visualization [24,25]; this need can be partially addressed via formal scientific models. If we accept that information visualization needs a formal foundation, the question remains whether the existing models from perceptual psychology and cognitive science are sufficient. The problem with these formalisms is they do not address the specific problems of visualization. While they provide general guidelines, models from non-visualization fields do not consider the context of the visualization environment - the user *and* the computer. What is needed is a set of formal foundations that bridges the gap between the general human experience and the visualization domain (Figure 1). We propose two models for this purpose: an *exploration* model that incorporates the user's interaction with the visualization and the dynamic aspects of their analysis, and a *transform design model* which encapsulates the depiction and constructive aspects of the visualization. These models would abstract fundamental principles of visualization science and design, and thus proscribe (via their predictive power) empirically driven practices.

4.3   Visualization Exploration Model

Visualization exploration is a goal-driven task incorporating visual search and information seeking. It is an iterative process - a user creates a visualization result, evaluates its worth, and then manipulates the visual parameters (e.g. color maps, selection regions) creating new results until satisfied. Thus, any formal model of computer-mediated visual exploration must capture the cognitive operations and how those realized actions manipulate the visualization. Cognitive operations are the domain of cognitive science, and several methods exist to model the human analysis process [26]. To bridge this work to visualization, two additional levels are needed: first, a description of the visual information search process and how it affects human cognition; second, a model of how the visualization session evolves due to human interaction. Visual sensemaking models such as the information foraging work of Pirolli and Card [27] begin to address the first need. Formalisms that capture the range of human-visualization interactions are targeted at the second [28,29,30].

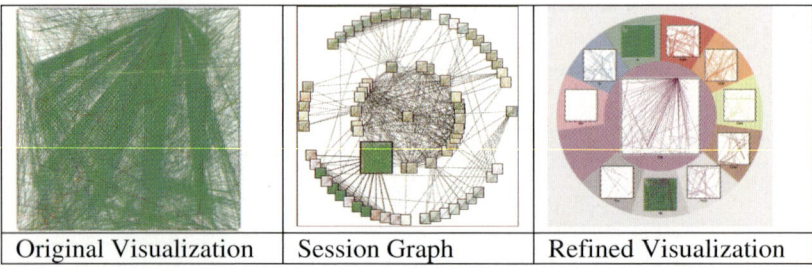

| Original Visualization | Session Graph | Refined Visualization |

Fig. 2. Analysis and evolution of a network traffic visualization. The original interface (a) uses colored lines connected to the edges of a square to depict changes. A formal model was used to capture interaction with the tool, and these sessions were analyzed to improve the interface (b). The redesigned interface (c) makes the exploration more efficient by displaying all event types individually and combined.

There are several benefits to a complete visualization exploration model. An understanding of how humans process visual cues in order to make exploration decisions can inform visualization design. For example, this "information scent" has been used to understand the cost-benefit trade-offs of different focus+context visualizations and to formally understand efficient web interfaces [31,32]. Similarly, a formalism for the visualization process lends itself to analysis to measure the user's efficiency of exploration [33,34]. Clusters of similar results (based upon metrics) during the session suggest redundant exploration; analysis of sessions based upon these metrics illuminate the path to more effective design [29,35] (Figure 2).

4.4   Visualization Transform Design Model

An exploration model describes and predicts a human's interaction with a visualization system based upon its design. This model neglects to describe the components that compose the design or provide initial design guidance. To provide this guidance, visualization transform design models are needed. A *visualization transform* is the function that computes the depicted result from visualization parameters – elements such as brushed graph nodes or opacity maps that dictate the rendered result. Significant work has expressed different mechanisms for constructing such transforms [36,37,38,39,40] (see Figure 3 for an extended example), but these categorizing efforts lack two things to provide a formal foundation. First, they do not utilize perceptual and cognitive literature to suggest and evaluate design decisions; second, most do not address the evolution of transforms and the utilization of extra-visualization tools important to the analysis (e.g., statistical packages). Efforts in providing design guidance for data display has been investigated [41,42,43] and formal algebras for transform modification have been recently presented [44,45]. However, these distinct contributions require effort to unify and to validate for a complete and cohesive scientific foundation.

A complete, predictive transform design model will yield several benefits. Toolkits for visualization creation benefit by providing guidance on suitable, less suitable, and unsuitable component choices. Visualization pedagogy will improve due to a validated foundation for techniques. Further, formal models will lead to objective metrics for evaluating a transform's effectiveness. All of these enhancements feed back into a visualization system, improving its potential utilization.

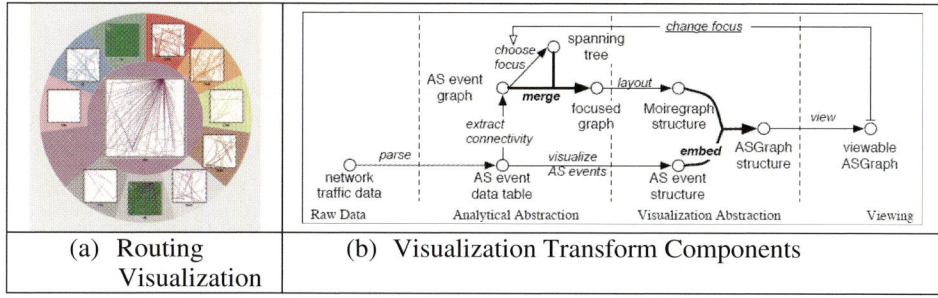

| (a) Routing Visualization | (b) Visualization Transform Components |

Fig. 3. A depiction of the revised network routing visualization transform. Nodes represent the state of the data (e.g., a table of events) while edges represent operators or interactions (e.g., parsing the data). In this example, the network visualization is combined with a graph visualization by embedding the results of the former within the latter. The graph itself is a composition, merging a spanning tree and the original graph to layout the selected sub-graph. Back-propagation of state due to interaction is included. The depiction is based upon an extended Data State transform model.

4.5   The Value of Visualization Craft

A research program investigating scientific grounding for visualization is not meant to diminish the importance of the engineering component of visualization. Visualization is a tool for humans; engineering efforts form the basis of providing such tools. Formal exploration and design models can guide the creation of a visualization system; however, as is the case now, multiple comparable techniques will often solve the same problem. Thus, the craft of visualization - the confidence in design choices gained through experience - will still be needed to decide between the choices a formal model provides. Inspiration and creativity will not be eliminated by a more rigorous foundation; the foundation will serve as a springboard for such endeavours.

4.6   What Is Left to Be Done

Formal foundations for a science of information visualization are still in a nascent stage. Elements of complete exploration and transform models exist; however, they have neither been reconciled with each other nor validated for their correctness. A close collaboration between perceptual psychologists, cognitive scientists, visualization researchers, and practitioners is needed to drive research into foundations: perceptual and cognitive scientists provide the human-based foundations, practitioners provide the case studies for observation and validation of models, and visualization researchers will form the domain-specific bridge between them.

Humans and computers play an integrated role in the development and utilization of visualization. A formal foundation would measure the efficiency of the former, and guide the design of the latter in order to create a more effective whole.

5   Conclusion

The three discussions of Information Visualization presented here draw on existing theories of data-centric prediction, information communication and scientific modeling, and relate in different ways to the linguistic framework defined in the introduction. A single uniting theory of Information Visualization may be impossible due to its strong relationship to and use of several other diverse disciplines (e.g. psychology (perception, cognition and learning), graphic design and aesthetics).

Investigating theoretical approaches used in other disciplines, and their relation to Information Visualization, is an obvious way forward, and can provide a useful way for researchers in the area to present, discuss and validate their ideas; it is hoped that the over-arching linguistics-based framework of representation, user exploration and manipulation, and system exploration and manipulation will prove useful in linking the constituent theories together. The more solid theoretical analyses that Information Visualization researchers or tool designers can call on in defending or validating their work, the more secure the discipline will be.

References

1. de Saussure, F.: Writings in General Linguistics. In: Bouquet, S., Engler, R., Sanders, C., Pires, M. (eds.), Oxford University Press, Oxford (2006)
2. Bakhtin, M.: The Dialogic Imagination, University of Texas Press (1981), quoted in Ball, A.F., Freedman, S.W.: Bhaktinian Persepectives on Language, Literacy, and Learning, Cambridge University Press, Cambridge (2004)
3. Andrienko, N., Andrienko, G.: Exploratory Analysis of Spatial and Temporal Data: A Systematic Approach. Springer, Heidelberg (2006)
4. Fayyad, U., Piatetsky-Shapiro, G., Smyth, P.: From data mining to knowledge discovery in databases. AI Magazine 17, 37–54 (1996)
5. Card, S.K., Moran, T.P., Newell, A.: The Psychology of Human-Computer Interaction. Erlbaum Associates, Hillsdale (1983)
6. Schneider, T.D.: Information Theory Primer. http://www.lecb.ncifcrf.gov/~toms/paper/primer (April 14, 2007)
7. Cherry, C.: On Human Communication, 2nd edn. MIT Press, Cambridge (1966)
8. MacKay, D.: Information, Mechanism and Meaning. MIT Press, Cambridge (1969)
9. Saraiya, P., North, C., Duka, K.: An evaluation of microarray visualization tools for biological insight. In: Proc. IEEE Symposium on Information Visualization, pp. 1–8 (2004)
10. Keller, P., Keller, M.: Visual cues: Practical Data Visualization. IEEE Computer Society Press, Los Alamitos (1993)
11. Cui, Q., Ward, M., Rundensteiner, E., Yang, J.: Measuring data abstraction quality in multiresolution visualization. In: Proc. IEEE Symposium on Information Visualization, pp. 709–716 (2006)
12. Bertin, J.: Matrix theory of graphics. Information Design 10(1), 5–19 (2001)
13. Seo, J., Shneiderman, B.: A rank-by-feature framework for interactive exploration of multidimensional data. In: Proc. IEEE Symposium on Information Visualization, pp. 96–113 (2005)
14. Peng, W., Ward, M., Rundensteiner, E.: Clutter reduction in multi-dimensional data visualization using dimension reordering. In: Proc. IEEE Symposium on Information Visualization, pp. 89–96 (2004)

15. Tufte, E.: The Visual Display of Quantitative Information. Computer Graphics Press, Cheshire (1983)
16. Ward, M., Theroux, K.: Perceptual benchmarking for multivariate data visualization. In: Proc. Dagstuhl Seminar on Scientific Visualization, pp. 314–328 (1997)
17. Fua, Y.-H., Ward, M., Rundensteiner, E.: Hierarchical parallel coordinates for exploration of large datasets. In: Proc. IEEE Conference on Visualization, pp. 43–50 (1999)
18. Novotny, M., Hauser, H.: Outlier-preserving focus+context visualization in parallel coordinates. IEEE Trans. Visualization and Computer Graphics 12, 893–900 (2006)
19. Dommik, G.: Do We Need Formal Education in Visualization? IEEE Computer Graphics and Applications 20(4), 16–19 (2000)
20. Borland, D., Taylor, R.M.: Rainbow Color Map (Still) Considered Harmful. IEEE Computer Graphics and Applications 27(2), 14–17 (2007)
21. Brewer, C.A.: Color use guidelines for data representation. In: Proceedings of the Section on Statistical Graphics, American Statistical Association, pp. 50–60 (1999)
22. Ware, C.: Information Visualization: Perception for Design. Morgan Kaufmann, San Francisco (2004)
23. van Wijk, J.J.: The Value of Visualization. In: Proceedings of IEEE Visualization, pp. 79–86. IEEE Computer Society Press, Los Alamitos (2005)
24. Johnson, C., Moorehead, R., Munzner, T., Pfister, H., Rheingans, P., Yoo, T.S.: NIH-NSF Visualization Research Challenges Report, 1st edn. IEEE Computer Society Press, Los Alamitos (2006)
25. Thomas, J.J., Cook, K.A.: Illuminating the Path: The Research and Development Agenda for Visual Analytics. IEEE Computer Society Press, Los Alamitos (2005)
26. Anderson, J.R.: Cognitive Psychology and its Implications, 6th edn. Worth (2005)
27. Pirolli, P., Card, S.K.: Information Foraging. Psychological Review 4, 643–674 (1999)
28. Brodlie, K., Poon, A., Wright, H., Brankin, L., Banecki, G., Gray, A.: Problem Solving Environment Integrating Computation And Visualization. In: Nielson, G.M., Bergeron, R.D. (eds.) Proceedings of the 4th IEEE Conference on Visualization, pp. 102–109 (1993)
29. Jankun-Kelly, T.J., Ma, K.-L., Gertz, M.: A Model and Framework for Visualization Exploration. IEEE Transactions on Visualization and Computer Graphics 13, 357–369 (2007)
30. Lee, J.P., Grinstein, G.G.: An Architecture For Retaining And Analyzing Visual Explorations Of Databases. In: Nielson, G.M., Silver, D. (eds.) Proceedings of the 6th IEEE Conference on Visualization, pp. 101–108 (1995)
31. Pirolli, P., Card, S.K., Van Der Wege, M.M.: Visual information foraging in a focus + context visualization. In: Proceedings of the SIGCHI conference on Human factors in computing systems, pp. 506–513. ACM Press, New York (2001)
32. Pirolli, P.: Rational Analyses of Information Foraging on the Web. Cognitive Science 29(3), 343–373 (2005)
33. Jankun-Kelly, T.J., Ma, K.-L., Gertz, M.: A Model for the Visualization Exploration Process. In: Moorhead, R.J., Gross, M., Joy, K.I. (eds.) Proceedings of the the 13th IEEE Conference on Visualization (Vis '02), pp. 323–330 (2002)
34. Lee, P.J.: A Systems and Process Model for Data Exploration. PhD thesis. University of Massachuesetts, Lowell (1998)
35. Teoh, S.T., Jankun-Kelly, T.J., Ma, K.-L., Wu, S.F.: Visual Data Analysis for Detecting Flaws and Intruders in Computer Network Systems. In: IEEE Computer Graphics and Applications, p. 24. IEEE Computer Society Press, Los Alamitos (2004)
36. Abram, G., Treinish, L.: An Extended Data-Flow Architecture For Data Analysis And Visualization. In: Nielson, G.M., Silver, D. (eds.) Proceedings of the IEEE Conference on Visualization 1995 (Vis '95), pp. 263–270. IEEE Computer Society Press, Los Alamitos (1995)
37. Chi, E.H., Riedl, J.T.: An Operator Interaction Framework For Visualization Systems. In: Dill, J., Wills, G. (eds.) Proceedings of the IEEE Symposium on Information Visualization, pp. 63–70 (1998)

38. Haber, R.B., McNabb, D.A.: Visualization Idioms: A Conceptual Model for Scientific Visualization Systems. In: Nielson, G.M., Shriver, B., Rosenblum, L. (eds.) Visualization in Scientific Computing, pp. 74–93. IEEE Computer Society Press, Los Alamitos (1990)
39. Hibbard, W.L., Dyer, C.R., Paul, B.E.: A Lattice Model for Data Display. In: Bergeron, R.D., Kaufman, A.E. (eds.) Proceedings of the 5th IEEE Conference on Visualization (Vis '94), pp. 310–317 (1994)
40. Schroeder, W.J., Martin, K.M., Lorensen, W.E.: The Design and Implementation of an Object-Oriented Toolkit for 3D Graphics and Visualization. In: Yagel, R., Nielson, G.M. (eds.) Proceedings of the 7th IEEE Conference on Visualization, pp. 93–100 (1996)
41. Casner, S.M.: Task-analytic approach to the automated design of graphic presentations. ACM Transactions on Graphics 10(2), 111–151 (1991)
42. Mackinlay, M.: Automating the Design of Graphical Presentations of Relational Information. ACM Transactions on Graphics 5(2), 110–141 (1986)
43. Roth, S.F., Mattis, J.: Data Characterization for Intelligent Graphics Presentation. In: Proceedings on Human Factors in Computing Systems (CHI'90), pp. 193–200 (1990)
44. Bavoli, L., Callahan, S.P., Crossno, P.J., Freire, J., Scheidegger, C.E., Silva, C.T., Vo, H.T.: VisTrails: Enabling Interactive Multiple-View Visualizations. In: Proceedings of the 16th IEEE Conference on Visualization (2005)
45. Weaver, C.: Building Highly-Coordinated Visualizations in Improvise. In: Proceedings 2004 IEEE Symposium on Information Visualization, pp. 159–166. IEEE Computer Society Press, Los Alamitos (2004)

Teaching Information Visualization

Andreas Kerren[1], John T. Stasko[2], and Jason Dykes[3]

[1] School of Mathematics and Systems Engineering, Växjö University,
Vejdes Plats 7, SE-351 95 Växjö, Sweden,
kerren@acm.org
[2] School of Interactive Computing & GVU Center, Georgia Institute of Technology,
85 5th St., NW, Atlanta, GA 30332-0760, USA,
stasko@cc.gatech.edu
[3] Department of Information Science, City University,
Northampton Square, London EC1V 0HB, UK,
jad7@soi.city.ac.uk

Abstract. Teaching InfoVis is a challenge because it is a new and growing field. This paper describes the results of a teaching survey based on the information given by the attendees of Dagstuhl Seminar 07221. It covers several aspects of offered InfoVis courses that range from different kinds of study materials to practical exercises. We have reproduced the discussion during the seminar and added our own experiences. We hope that this paper can serve as an interesting and helpful source for current and future InfoVis teachers.

1 Introduction

Education is an important aspect of any emerging and rapidly evolving discipline and this is certainly the case in Information Visualization (InfoVis) with its emphasis on the exploratory development of knowledge. Most of the researchers participating in the Dagstuhl seminar and contributing to this volume are involved in helping students graduate with competencies in visualization. The growing number of courses in Information Visualization is matched by the variety of styles of courses offered, in terms of course content, materials used, and evaluation methodologies. Attendees at Dagstuhl seminar were curious to learn about the courses others offered and the approaches and resources that were being used, and so a session on Information Visualization teaching and education was held.

To prepare for that session and benchmark current offerings, Keith Andrews from Graz University, Austria, prepared a survey about InfoVis-related courses and distributed it to the attendees. The survey was intended to gather a variety of information, mostly demographic, including teaching styles, textbooks, enrollments, teaching aids, examinations, etc. Nineteen participants completed the survey and described their courses. This paper presents the survey results and includes the perspectives of some of the participants in relation to their own teaching experience in light of these and discussions amongst colleagues at Dagstuhl.

A. Kerren et al. (Eds.): Information Visualization, LNCS 4950, pp. 65–91, 2008.

The survey consisted of four different parts. The specific questions included in each part are listed below.

1. General Information
   (a) Instructor name
   (b) Educational organization
   (c) Title of course
   (d) Course home page (URL)
   (e) Last taught (date)
   (f) Course level (graduate or undergraduate)
   (g) Course hours per week
   (h) Course number of weeks
   (i) Enrollment (number of students)
2. Teaching Aids
   (a) Do you use one or more textbooks (yes, no)?
       i. If so, which ones?
   (b) Do you assign papers for compulsory assigned reading (yes, no)?
       i. If so, which ones?
   (c) Do you have your own set of lecture notes (yes, no)?
       i. URL (if available)?
   (d) Do you have teaching assistants for the course (yes, no)?
       i. If so, how many?
3. Practical Exercises (Projects)
   (a) Do you use practical exercises or projects (yes, no)?
       i. If so, please describe a typical exercise or project.
       ii. If so, how do you grade the practical exercises or project?
4. Examination or Test
   (a) Do you have an examination (yes, no)?
       i. If so, written or oral exam?
       ii. If so, please describe a typical exam question.

Firstly, we briefly review the results of the survey. Section 3 summarizes the topics discussed during the interactive session on teaching at the seminar. Finally a selection of participants reflect upon how these issues relate to their own experience of teaching Information Visualization in Section 4.

2   Results

We present the results in four sections, one for each of the sections of the survey.

2.1 General Information

The first part of the survey gives an overview of the courses offered at the different universities represented by the Dagstuhl participants and provides some details about the courses themselves. Table 1 shows the responses obtained from this part. A balance of European and North-American universities were represented by participants in the survey results. The majority of courses were focused on the "core field" of information visualization (about 68%). Two courses were about visualization/computer graphics in general, and the rest were about application fields (e.g. geographic visualization) or broader topics, such as information interfaces or visual communication. This scope reflects the broad and interdisciplinary nature of Information Visualization and provides some indications as to why developing an agreed Information Visualization curriculum may be difficult.

Most of the courses (79%) had their own publicly accessible web page providing access to course related information. Nearly all the referenced courses were given in 2006 and 2007. Since all the responding instructors are active researchers in the field as well, we can assume that all these courses covered the current state of the art in information visualization. Because the detailed curriculum for the courses was not part of the survey, we do not have details about actual course content. The web pages associated with each of the courses are a rich source of information however and we used these to gather keywords associated with the curricula of each. Figure 1 shows a tag cloud generated from these keywords that gives a flavor of the variety and importance of different topics across the courses. The dominant words reflect some of the tensions in Information Visualization education, with a collective need to focus on data—its dimensionality and structure, techniques for layout and visual encoding and people and their responses to these methods and the systems through which they are accessed. Perhaps the tag cloud and the varied responses suggest a need for systematic research to learn about the range of approaches that are used in teaching Information Visualization and related topics. The session discussion, summarized in Section 3, led to more insight about this, but it was not a comprehensive examination.

Most courses were taught at the graduate level, with only two being undergraduate courses. At the bottom of Table 1, descriptive statistics about the results of questions Q1g-Q1i on course duration and size are provided. The average duration of a course and the number of hours of weekly meeting time are relatively consistent across the group. Note that most class sizes are relatively small, echoing the fact that Information Visualization is still a relatively new and growing area. Here, the undergraduate course #16 seems to be an outlier because of its large enrollment. However, this course is a compulsory course on computer graphics and visualization, and the instructor plans to divide this course into two parts in the future.

2.2 Study Materials and Teaching Aids

Textbooks (Q2a): About 72% of all the instructors (13 in total) used one or more textbooks in their courses. The most popular books were those by Colin

Table 1. Results of Part 1 of the Survey on General Information together with a brief descriptive Analysis.

#	Q1a	Q1b	Q1c	Q1d	Q1e	Q1f	Q1g	Q1h	Q1i
1	Keith Andrews	TU Graz, Austria	Information Visualisation	[4]	Summer 2007	g	3	14	15
2	Jason Dykes	City University London, UK	GeoVisualization	–	Spring 2007	g	3	12	22
3	Achim Ebert	TU Kaiserslautern, Germany	Information Visualization	–	Winter 04/05	g	2	14	18
4	Helwig Hauser	TU Vienna, Austria	Information Visualization	[23]	Summer 2007	g	3	15	30
5	Jeffrey Heer	UC Berkeley, USA	Visualization	[25]	Spring 2006	g	3	16	20
6	T.J. Jankun-Kelly	Mississippi State University, USA	Information Visualization	[28]	Fall 2006	g	3	15	12
7	Daniel Keim	University of Constance, Germany	Information Visualization	[29]	Summer 2007	g	5	14	20
8	Andreas Kerren	TU Kaiserslautern, Germany	Information Visualization	[34]	Winter 06/07	g	2	14	9
9	Robert Kosara	UNC Charlotte, USA	Visual Communication in Computer Graphics and Art	[39]	Spring 2007	g	3	18	15
10	Kwan-Liu Ma	UC Davis, USA	Information Visualization	[43]	Winter 2006	g	3	10	12
11	Kwan-Liu Ma	UC Davis, USA	Information Interfaces	[44]	Spring 2007	ug	3	10	17
12	Guy Melançon	University of Bordeaux, France	Models and Algorithms for Information Visualization and Bioinformatics	–	Fall 2006	g	3	11	15
13	Silvia Miksch	TU Vienna, Austria	Information Visualization	[48]	Winter 2006	g	3	14	30
14	Tamara Munzner	UBC Vancouver, Canada	Information Visualization	[51]	Fall 2006	g	3	13	15
15	Chris North	Virginia Tech., USA	Information Visualization	[53]	Spring 2007	g	3	15	25
16	Jonathan Roberts	University of Kent, UK	Computer Graphics and Visualisation	–	Summer 2005	ug	2	12	60
17	John Stasko	Georgia Tech., USA	Information Visualization	[64]	Spring 2006	g	3	15	35
18	Matt Ward	WPI, USA	Data Visualization	[70]	Spring 2006	g	3	14	10
19	Jing Yang	UNC Charlotte, USA	Information Visualization	[72]	Spring 2007	g	3	11	9
					Arithmetic Mean		3	14	20
					Median		3	14	17
					Standard Deviation		0.6	2.1	12.1

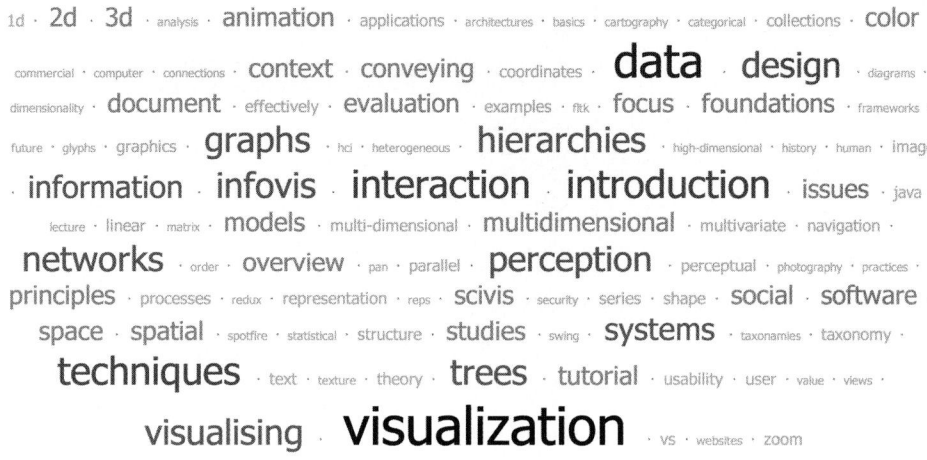

Fig. 1. Tag cloud of course topics. Includes courses with Web pages in the English language for which a URL was provided.

Ware and Robert Spence. The following list shows all books used by instructors of InfoVis courses in descending order of popularity.

1. Information Visualization: Perception for Design. Colin Ware [71].
2. Information Visualization: Design for Interaction. Robert Spence [62].
3. Readings in Information Visualization: Using Vision to Think. Stuart Card, Jock Mackinlay, and Ben Shneiderman (Eds.) [8].
4. Envisioning Information. Edward Tufte [67].
5. Visualisierung – Grundlagen und allgemeine Methoden, Heidrun Schumann and Wolfgang Müller [58] (in German).
6. – Human-Centered Visualization Environments. Andreas Kerren, Achim Ebert, and Jörg Meyer (Eds.) [35].
   – Information Visualization: Beyond the Horizon. Chaomei Chen [9].

Some respondents noted that they also used Tufte's other books [69, 68] for specific aspects of the course or as a focused topic, rather than as a general textbook. Other courses cover more general fields, such as (Data) Visualization (#5, #18, . . .), with information visualization as part of them. In these courses, other textbooks were used, for example the VTK Book [57], *Designing Visual Interfaces* by Mullet and Sano [50], or [58, 30, 42, 16, 18]. Those teaching visualization in a particular domain (e.g. GeoVisualization (#2)) used more specific texts associated with the relevant discipline [60, 15, 49]. Figure 2 shows the usage of books for all courses listed in Table 1.

Papers for Compulsory Assigned Reading (Q2b): 68% of the courses used research papers for compulsory assigned reading. Many different papers were used so an exhaustive listing here is not appropriate. Most papers were

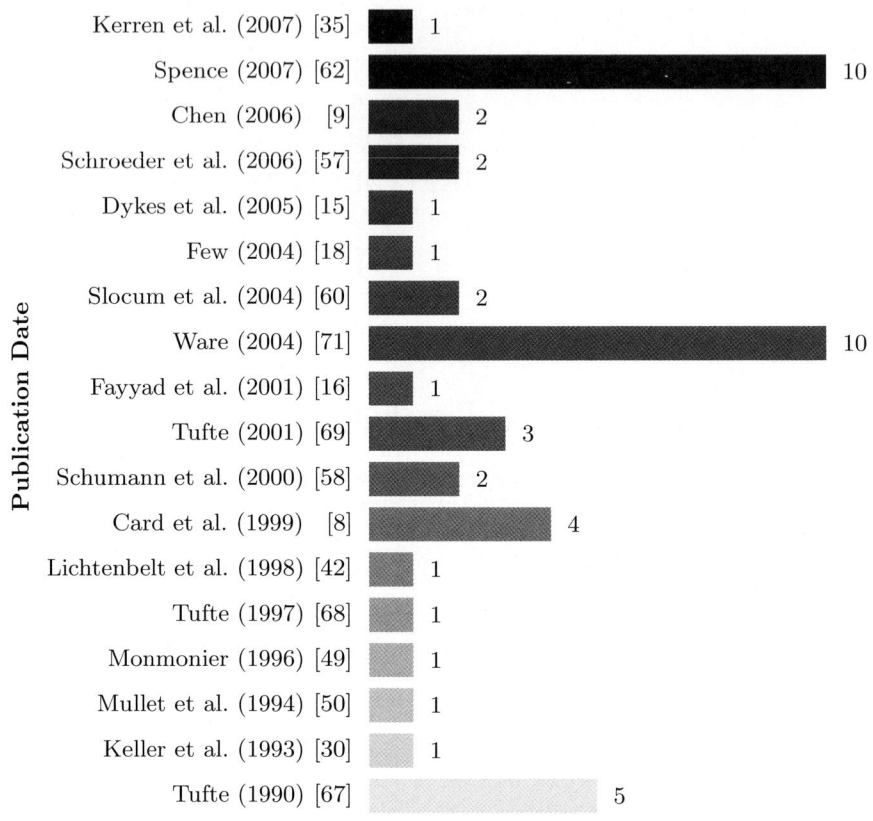

Absolute Usage (13 instructors in total)

Fig. 2. Usage of books for all courses ordered by publication date.

selected from the IEEE InfoVis and Vis Conference proceedings as well as from the ACM CHI proceedings.

Students typically had to prepare a short presentation about a research paper in these courses. This helps students gain skills in oral communication (particularly if presentations are critiqued in class) and helps the courses to explore a variety of different visualization approaches and techniques in discussion. Such a presentation also could be part of a larger practical exercise or project (see Section 2.3).

Own Lecture Notes (Q2c): Interestingly, all the instructors used their own course lecture notes (many as PowerPoint slides). 58% published their lecture notes on the course web site without any restriction. We assume that the remaining instructors either offered their notes on the web with restricted access or simply used the notes to lecture from.

Teaching Assistants (Q2d): More than the half of all instructors (58%) had no teaching assistant (TA) to support their course. In these cases, we can assume that instructors also supervised practical exercises which were offered in almost all courses. Only two courses were supported by two TAs, and the rest had one TA. These figures may be due to small classes that were reported in most cases, but might be indicative of a lack of support for teaching and learning in Information Visualization, which is a very practical activity. Any such trend may be of concern to InfoVis educators.

2.3   Practical Exercises

Survey analysis showed that nearly all courses (95%) had practical exercises or projects. However, a wide variety of different types of exercises were noted. One of the most common practical exercises cited was to have students use InfoVis tools and then critique them. For example, students may choose from a set of provided interesting datasets (election results, finance data, etc.) and then they examine the datasets using the different InfoVis systems to reason about analytic questions or tasks provided by the instructor or generated by the student. These exercises often conclude with a written report evaluating the tools and their effectiveness for the analysis. Alternately, some instructors give students freedom to choose/generate their own input data. From the survey results, the tools used most in such exercises are as follows:

- *Spotfire*, TIBCO Software, Inc. [63],
- *TableLens*, Inxight Software, Inc. [66],
- *ILOG Visualization Suite*, ILOG, Inc. [26],
- *Tableau Desktop*, Tableau Software, Inc. [65],
- *InfoZoom*, humanIT Software GmbH [27], and
- *KidPad*, University of Maryland [38].

The KidPad tool provides a set of zooming user interface and visualization facilities so it can be used to develop small stand-alone visualizations on individual topics. Thus, students may be asked to build a visualization on the basis of self-chosen or self-generated data sets together with some conditions, e.g., to think about designing overviews or spatial layout.

Another common approach of practical exercises is based on the idea that students should implement the fundamental idea of a research paper. As a first step, students (typically groups of at most three students) choose a paper from a list given by the instructor. Next, the students prepare a brief presentation to explain the fundamental idea and to give an overview on the planned implementation. This presentation helps to minimize the danger that the students will become too focused on low-level details. To conclude, the students present and demonstrate their tools in the classroom.

A last example of a common practical exercise is based on a special type of data in information visualization: graphs and networks. In order to help students better appreciate the difficulty of graph layout, the instructor has students draw a small graph of 10-15 nodes based on only the connectivity of the graph's vertices. Students are instructed to make their graph drawing as aesthetically pleasing as possible. When all the drawings are submitted, students view all the different efforts and vote to select the best ones. In class discussion, students explain why they voted as they did, and this can lead to a discussion of graph drawing metrics, e.g., that edge crossings and bends have a negative effect on aesthetic quality, etc. Such metrics are the basis of graph drawing algorithms that could be analyzed by the students in a second step.

Grading of Practical Exercises: One can classify the described approaches for practical exercises into two main groups:

1. implementation of a specific technique or implementation of a visualization to solve a specific problem, and
2. examination of the usefulness of a given tool or visualization approach, such as a commercial visualization system or graph drawing metrics, etc.

Grading of assignments of the first type can based on the overall quality of the implementation itself, i.e., the instructor and/or TA evaluate the needed time, aesthetic aspects, level of effort (complexity), usability, and capability provided by the tool to get more insight about the chosen data set. Both types of assignments can be combined with an oral presentation in the classroom. In this case, the quality of the presented slides, lecture style, originality, and the presentation quality itself can be used to evaluate presentation skills if these are intended learning outcomes—and it is increasingly acknowledge as desirable to incorporate the teaching and learning of skills into the subject-area curriculum rather than dealing with this independently. The survey yielded one interesting case in which the students themselves voted on the best presentation(s).

2.4   Examination

More than the half (63%) of the surveyed courses had examinations of some kind. The most common form was a written exam (37%), particularly in the United States, but a notable portion (21%) used oral exams and 5% used both. Oral examinations were used in Europe only, where there is a tradition of oral examination for advanced level courses.

The survey also gave some insight into the different kinds of typical exam questions (Q4a[ii]). There are a lot of different variants; a selection of the most asked questions includes:

– Explain technique X for the visualization of problem Y.
– Given is a concrete problem and a task to be fulfilled. Which technique would you use?
– Compare technique X with technique Y.
– Explain the construction of a Treemap, Starplot, Circle Segments, ...
– What are the advantages/disadvantages of technique X?
– What is a preattentive feature?
– What are the principles of using color?

This list only gives a rough overview about the issues that are important for the instructors. The course web pages do not provide additional detail—we found no specific exam questions or model answers when investigating methods of examination further. In general, we could observe that examiners focus not only on technical approaches or methods, but also students' capabilities for critical reflection and to demonstrate their working knowledge of human visual perception.

3   Seminar Discussions

In the seminar session about teaching, Keith Andrews first presented the initial survey result data. Next, workshop attendees discussed a variety of issues related to teaching including "best practices" and ways to improve all our courses.

Curriculum:: A common problem reported by the attendees was some uncertainty about how to logically organize the set of topics in an InfoVis course. For example, Robert Spence's 2nd edition textbook on information visualization follows the classical pipeline model of representation (data types, tree representations, ...), presentation (space and time limitations, including zooming, distortion, ...), and interaction (navigation, browsing, mental models, ...) [62]. Many alternative ways of organizing course content exist, however. One suggestion was to consider four cross-cutting dimensions: data types, domains, techniques, and methodologies. Cognitive and perceptual issues were presented as an alternative dimension of importance. The group did not come to any decisions about what the most suitable structure would be since this is clearly dependent on the orientation of the course and its learning aims. Simply being aware of alternate orderings is valuable as instructors consider alternatives however.

Study Materials: The discussion on the use of research papers for compulsory reading identified many different strategies for doing so. Many attendees echoed a frustration about the difficulty in getting students to actually read assigned articles, so many of the strategies addressed this particular issue. Several participants reported about their own experiences and ideas, a number of which are listed below.

- Papers are assigned, and students must present them. This takes place in parallel with the regular lectures. This procedure seems to be pedagogically beneficial because students learn to read actual research work, to prepare a short talk and to give a presentation in the classroom. A disadvantage is that student presentations vary greatly in quality. Some colleagues reported on students losing interest in and not learning from poor presentations—they would prefer the instructor to do all the lecturing. It is unclear if this is truly a disadvantage, however. Perhaps, the amortized learning benefit is high enough and this would justify the approach.

- The instructor gives lectures on mandatory meetings. Students can pick specific topics and lecture about a topic that the other students have not read about.
- Papers are assigned, and there are written/oral questions on readings.
- Papers are assigned, and students must write a structured critical review (about half a page) of them, i.e., a paragraph on the paper's content, an evaluative paragraph and an indication as to whether and why other students might read the paper.

Practical Exercises: Software projects or practical exercises were viewed as being very important to most attendees. These projects allow students to demonstrate/learn the difficulty of many practical problems, such as the drawing of a graph or navigation in large information spaces. Using a visualization tool (Spotfire, TableLens, ...) allows the students to interact with different visual representations and to gain experience about the advantages/disadvantages of different visualization techniques. Furthermore, students become acquainted with commercial tools. One interesting experience of attendees following this approach was that students' negative impressions of InfoVis systems mostly involved user interface or HCI issues, not the actual visualization technique(s).

One problem with such exercises is the potential difficulty in gaining commercial software for use in the projects. More vendors are making their systems freely available for educational use, however. Tableau Software is an example of a company doing so. Another issue is the challenge of finding "good" data sets to be used by the students. One suggestion was use data sets from previous InfoVis Conference contests but most attendees felt that the contest datasets are too big and complex for introductory courses. Still, the contest data sets may be suitable for advanced level courses where students have a good background in the most important techniques. Perhaps, the contest should include the production of a data subset, specifically designed for educational purposes.

Using Other Media: Workshop attendees discussed that a large and comprehensive public collection of InfoVis-related images and videos would be very helpful for instructors. Videos of interaction scenarios that show the usability and interaction capabilities of the tools would be especially beneficial. Images also could help to illuminate the history of InfoVis and illustrate different visualization techniques. Unfortunately, gathering a collection of images or videos in this way could cause copyright problems. This may be why many instructors have their own image/video archives with private access. The HCC Digital Library [24] of Georgia Tech is an example of an effort to gather a large collection of educational resources, but it is focused broadly on HCI, not just InfoVis.

Another possibility to obtain video material is to examine conference DVDs, such as the annual VIS/InfoVis/VAST DVD. Many contributions provide an additional video to clarify the usage and interaction techniques of their work. Again, it may be beneficial to encourage attendees to develop video summaries of their work, specifically for teaching and learning.

4 Personal Perspectives

In this section, three of the workshop attendees provide their own unique perspective on teaching InfoVis and InfoVis-related topics.

4.1 John T. Stasko, Georgia Institute of Technology, USA

I have been teaching a graduate course on Information Visualization at Georgia Tech since 1999. My university changed from a quarter system to a semester system then and I decided to create a course for this area that was becoming my research focus and growing in interest worldwide. Over the years I have thought of this course not merely as a teaching assignment but as a fundamental part of my academic portfolio and research mission. The course provides training for students to learn about the area and do subsequent research with me, or simply to apply their knowledge in business or government. Also, the course has directly led to a number of the research contributions made by students in the course and by my research group. Student projects from the course have won major contests [20, 54] or led to research papers [12]. Additionally, my dissatisfaction with the state of knowledge and background articles on particular course topics led to projects undertaken by my research group in those areas (i.e., analytic goals [3], user tasks [1], and interaction [73]). Below I will provide more details about the course and the projects that have resulted from it.

The Information Visualization course (CS 7450) is usually offered in the Spring semester each year. At Georgia Tech, semesters are 15 weeks long and and my course meets for 1.5 hours twice a week. The web pages for the most recent course offering can be found at `http://www.cc.gatech.edu/~stasko/7450`.

Human-computer interaction (HCI) is an important area of research in my home School—we have both a Masters degree in HCI and a doctorate in Human-Centered Computing (HCC), in addition to our undergraduate and graduate degrees in computer science. Students are drawn to the HCI and HCC degree programs from a variety of undergraduate majors, so many do not have formal computer science training. I made a conscious decision to designate the graduate HCI course as the only prerequisite for Information Visualization in order to encourage students from a wide variety of disciplines to enroll. Consequently, many students who do not have a strong background in programming typically take Information Visualization.

This fact has implications on the way that I teach the course. I do not use homework assignments in which the students must implement a visualization technique or system. Instead, assignments are more oriented toward design, critiquing and evaluation. I employ a group project where I do require that some type of software system be built, but the course demographics allow there to be at least one or two team members who are experienced programmers.

My high-level goal for the course is to have students learn the different information visualization techniques that have been created including the strengths and weaknesses of each, and to have the students become better critics of information graphics and visual systems. More specifically, the learning outcomes for the course include

- Students should gain an in-depth understanding of the field of Information Visualization including key concepts and techniques.
- Students should be able to critique visualization designs and make suggestions to improve them.
- Students should be able to design effective visualization solutions given new problems and data domains.
- Students should learn about the spectrum of commercial system solutions available in this area and how to choose one for a particular task or problem.

Perhaps the main challenge that I have faced in this course over the years is to construct a coherent syllabus and flow of topics throughout the term. Information Visualization is still a new area that is growing and maturing. Consequently, it does not exhibit a well-understood and agreed-upon set of topics that flow smoothly from one to the next. In my experience teaching the course, a number of key ideas have risen to the surface and I make these the important components of the course:

- **Data foundations** - A description and model of the different types of data that are encountered and how this data is transformed and stored for easier subsequent manipulation.
- **Cognitive issues** - A discussion of the user's goals and tasks in using an information visualization system. What cognitive benefits can visualization provide?
- **Visualization techniques** - A description of the different visual representations and interaction techniques that have been invented.
- **Interaction** - A discussion of the different types and the many issues surrounding interaction.
- **Data types/structures** - An introduction to specific types of data (e.g., time series, hierarchical, textual) and the visualization techniques that are well-suited at representing those data types.
- **Data domains** - An examination of different domains (e.g., software engineering, social computing, finance and business) and the visualization techniques that are helpful to people working in those areas.
- **Evaluation** - A dialog about the challenges of evaluation in information visualization and a review of different evaluation techniques that have been used in the area.

Some of these topics are fundamentally interwoven so the flow of concepts is not clearly self-contained and independent. For instance, certain visualization techniques are best used for specific data types (e.g., treemaps for hierarchical data). In organizing the course content, I feel this tension and often struggle with which topics to teach first. Nonetheless, my course uses this progression of topics as its organizational framework.

The course is lecture-based but I try to engage the students in discussions about the different concepts being studied. I have used Bob Spence's textbook some terms augmented by selected papers, and other terms I have used only research papers. I have settled on having students read one or possibly two research papers for each class. Typically, the paper is an important one for that topic or it is a good overview of the issues involved. When I have assigned more papers than this, I find that the students often do not adequately prepare and read all the papers. To cover more recent research, I typically select two or three recent articles on the topic of the day and I assign two or three students who must recap and describe their particular paper's key ideas to the class in less than five minutes. I believe that experience giving presentations like this is important and valuable to the students. All of my lecture slides can be found at the course website and in 2007 I created in-studio video versions of each lecture. These videos can be found at the website `http://vadl.cc.gatech.edu`.

I use a number of relatively small homework assignments in the course, along with one larger homework and a group project. I will frequently employ a midterm or final exam as well. The small homeworks often involve a visualization design exercise (on paper) given a data set. Of course, such assignments do not engage the interactive component of information visualization that is so important, so they are fundamentally limited.

The larger homework assignment is a commercial tools critique. Students are given five example datasets and asked to choose the two that they find most interesting. Before using any systems, the students examine the datasets and generate questions about them. Next, the students use a few information visualization systems to explore the data and try to answer those questions. I also alert the students to note any serendipitous findings that occur during exploration. Finally, the students must write a report in which they critique the different systems used, the visualization techniques each employs, and whether the systems led to insights and discoveries. I have used systems such as Spotfire, SeeIt, Advizor, Eureka (Table Lens), InfoZoom, InfoScope, and Grokker over the years. I find this assignment to be extremely valuable to the students as it allows them to gain hands-on experience with sophisticated systems and shows them how visualizations can (or cannot) be helpful in analysis and exploration.

This particular assignment even led to an interesting research contribution by my group. We studied the analytic queries generated by students over many years of the course and clustered these inquiries into different low-level analytic tasks that visualizations may assist. Our taxonomy of these tasks was presented at the 2005 Symposium on InfoVis [1].

I also employ a group project in the course in which students design and build a visualization system for a particular problem and data set. Teams of three or four students work together for most of the term and find a client with a data analysis problem or they simply choose a data set and envision the kinds of analytic queries that one would expect on it. The students explore different visualization designs, then they choose one to implement. In the past, student teams have often chosen to work on the contest datasets from the IEEE InfoVis or VAST Conferences. In fact, student teams from my course have won these contests on multiple occasions or have had competitive entries in the contests [20,54]. Group projects have even led to full papers at the InfoVis Symposium as well [12].

One ongoing tension with the group project is simply when to begin the assignment. By initiating the project early in the term, students have more time to work on it and make better progress. However, at that early point, students have engaged very little course material and so their understanding of information visualization concepts and ideas is not as rich. I have found that the simple topic chosen for the project can have a profound impact on the results, and better knowledge of the information visualization area leads students to make better choices in project topics. This has led me to wait until the midterm point to distribute the project in some semesters, but then the students have much less time to work on it.

Overall, the Information Visualization course has been valuable to me in many different ways. Perhaps most importantly, the process of preparing lectures and course material has made me reflect on the topics that I would be discussing and question "accepted" knowledge in the domain. I believe that this has made me a better researcher and it has generated ideas for new projects and investigations.

4.2   Andreas Kerren, Växjö University, Sweden

My experiences in teaching Information Visualization go back to the year 2003. At this time, I was a temporary assistant professor at the Institute of Computer Graphics and Algorithms of the Vienna University of Technology, Austria. The institute offered one course on InfoVis and several other related courses. In this environment, I had the opportunity to give several lectures on Software Visualization and domain-specific visualization, such as visualization in Bioinformatics. In 2005, I moved to the University of Kaiserslautern in Germany. There, I was responsible for the annual InfoVis courses. Based on my experiences from Vienna and current flows in research, I designed a completely new syllabus for this course. Originally, this syllabus provided 15 lectures (one semester at TU Kaiserslautern) plus practical exercises for Masters level students; each lecture took 1.5 hours once a week. As I have been appointed for a faculty position at Växjö University (VXU), Sweden, in 2007, some modifications were needed to address the different course and teaching system at this university. Web pages (in English) for the most recent InfoVis courses at VXU can be found at `http://cs.msi.vxu.se/isovis/courses/`. In this section however, I will focus to my experiences with my courses given at TU Kaiserslautern from 2005-2007, because the details of my last course there (in the winter semester 2006/2007 (WS06/07)) reflect my particulars within the Dagstuhl survey.

My general learning aims for the course are more or less identical to John Stasko's four learning outcomes at Page 76. Therefore, I don't want to repeat them at this place. It was very important for me to give students the opportunity for critical reflections and to show the newest directions in research. Furthermore, my course covers basic principles from cognitive psychology that have influence on InfoVis, such as human visual perception or Gestalt laws. As a result of this position, each technical approach and tool was discussed with respect to its value (please compare [17] in this book), its usefulness, and—if existent—its usage in

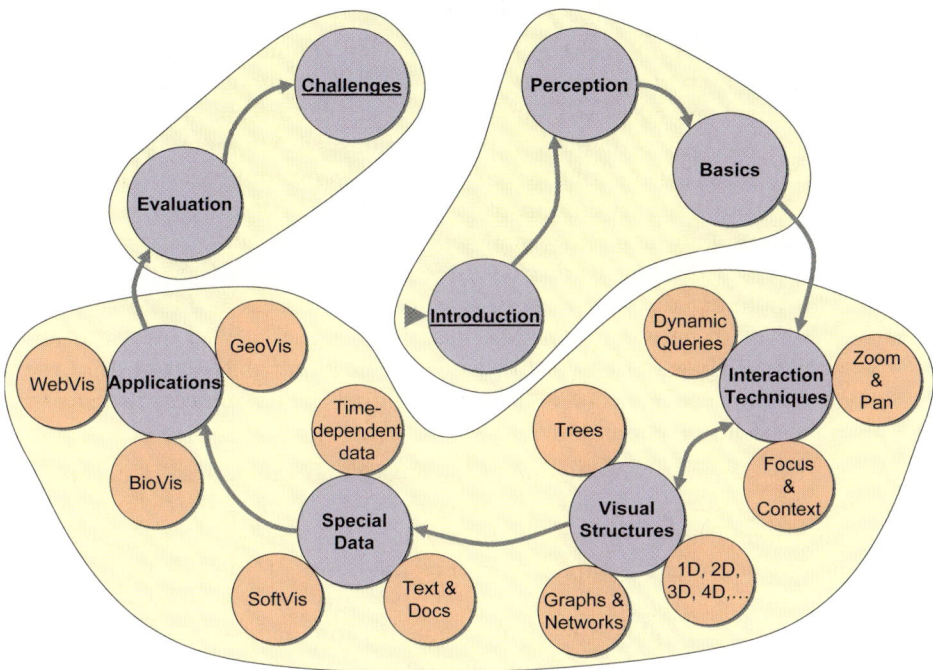

Fig. 3. Course structure of the WS06/07 InfoVis course given by Andreas Kerren at TU Kaiserslautern.

commercial products. Of course, this was a challenge and sometimes a little bit subjective because of missing quality metrics or missing evaluations of tools and techniques.

The design of the syllabus was partly influenced by courses given at Georgia Tech and TU Vienna in 2005 as well as by the textbooks of Spence [61], Ware [71], a pre-version of the textbook of Kerren et al. [35], and many research papers. The course is divided into three parts which are illuminated by Figure 3:

1. In the first part, I discuss basic knowledge that is important for the design or analysis of InfoVis concepts. As a first step, I introduce the field itself, give motivations for the need, and present several traditional and modern examples, mainly from Tufte's [67, 69] and Spence' books. Important is the differentiation between InfoVis and SciVis, also, if that is not so easy in some cases. After this introduction, a larger discussion on perception and cognitive issues is given. Here, I provide information about the perception of colors, textures, etc., preattentive features, and Gestalt laws. This course component is mainly based on the book of Ware, but also on a lot of examples and animations that can be found in the WWW. The last lecture of this first part describes basics, such as the InfoVis Reference Model (data tables, visual mapping, interaction, etc.) or data types and dimensionality. These issues are mostly based on the book of Card et al. [8].

2. The second part is the largest one of my course. Here, I discuss the most important interaction techniques at first, for example Dynamic Queries, Zoom&Pan, and several Focus&Context related techniques. This component is more or less geared to the InfoVis Reference Model [8], i.e., I distinguish interaction by means of data transformations, visual mapping, and view transformations. I use actual research papers to exemplify the different approaches. From a didactic point of view, this is a little bit tricky, because I presume some knowledge in visual representations or structures to explain my examples. I decided to discuss interaction at first and the visual structures for different data types after this. One advantage is the possibility to refer later to discussed interaction techniques directly with less additional explanations. My experiences with students show that they accept this order, and that they have no problem in understanding the differences/correlations. But this should be communicated previously.

   As described before, a discussion of the most important visual structures for more basic data types follows the interaction component. Here, I introduce visualization techniques for multivariate data, hierarchies and graphs mostly on the basis of research papers as well as the books of Spence and Kerren et al. Individual solutions for special kind of data types, e.g., time-series data, text, or software, follow this component directly. The second part finishes with visualization techniques for different data domains, such as BioVis, WebVis, GeoVis, etc. During the past years, I vary this part of the lecture a little bit depending on my current research interests or hot-topics. Resources for these lectures are current papers and articles, but also the second part of the textbook [35].

   Each course component of this second part is accompanied by short video or tool demonstrations. From my perspective, this is absolutely needed, especially for the different interaction techniques and their interplay with visual structures. It is fun to keep an eye on the students during such demonstrations, and as a result they are motivated to ask deeper questions. One interesting and traditional example is the claim to preserve the mental map in dynamic graph drawing. Only with the help of a video or demo it is possible to illustrate the difference between morphing or other techniques, such as foresighted layout [14]. However, the usage of video or tools is not always possible because of unavailability.

3. My course concludes with 1-2 lectures on possible evaluation techniques and the most important InfoVis challenges for the next years. Because of the missing time at the end of the semester, I focus on specific aspects of these issues. For example, the intended learning aim is to impart students an overview knowledge of basic evaluation techniques and—perhaps more important—an impression of the difficulties to perform such an evaluation. The final discussion of the most important challenges gives an idea about the current state of the field and leads to take part of it, for example, by working on a thesis in my research group. A good source for these issues are the corresponding chapters in Kerren et al. [35].

Another important part of the course are the assignments. They are composed of a brief presentation, of a software implementation, and of a short software demonstration at the end of the semester. Each student or student group (consisting of maximal two students) chooses a specific research paper from a list given on the course web page. I take care that the papers' topics and the presented approaches are not too complex. The final aim of the assignment is that the *main idea* of a paper should be implemented in any programming language. It is not needed that all features or interaction possibilities are implemented. But a GUI is mandatory in order to give me and the other students the chance to load another input file etc. Data sets depend on the paper topic, e.g., if the paper presents a new treemap layout then the students can choose their own input, such as the hierarchical file system on their own personal computer. All these topics should be discussed in the first presentation in the middle of the course. In a first step, each student or group prepares a presentation (10 minutes plus 3-5 minutes of discussion) about the chosen paper followed by a working plan. In this way, I can steer the processes, give hints, and prevent nasty surprises. At the end of the course, all implementations are presented and discussed in class. I have found that this division into two presentations and demos respectively helps students to think about important concepts. Furthermore, they have already learned the most important theoretical concepts during the course before they start to program. My overall impression of this practical project is very positive. At the beginning, the students often had doubts because it appears time-consuming and complex, but they had a lot of fun in the progress of the semester. The results were mostly really great; for me it is important that they learn to see the difficulties and to carefully reflect about the paper, not so much the result itself. Often, however, the resulting programs were amazingly good. My pedagogical concept, especially for the assignments, clearly follows moderate constructivistic learning approaches, as described in the following written by Jason Dykes or in some of my papers on learning concepts in context of using Software Visualization techniques [33, 32, 59].

The course evaluation by my students led to very good results for this course. They liked the way I structured the course, the motivating examples and videos, and they had the subjective feeling that they have learned a lot of interesting things. At large, it was not difficult to motivate students for InfoVis. It is a very interesting field also suitable for the solution of practical problems. Therefore, it was sometimes not so easy to explain why people cannot find more InfoVis in standard software products. This leads to a problem that is discussed in paper *The Value of Information Visualization* [17] of this book.

I would like to discuss one further issue that is important to me: Finding a good balance between giving a good overview of the field as compared to explaining the details of specific visualization techniques is pretty difficult, especially in the frame of 15 course lectures. Some students liked to get more overview knowledge of InfoVis, but they also disliked that some topics were only briefly covered. For instance, I used 1-2 lectures for the visualization of graphs. It is enough time to explain the most important things, but not enough time to explain the different graph drawing techniques in detail. Thus, I abstracted in many cases, but some students would like to learn more. The level of detail/abstraction in teaching InfoVis is not obvious. In general, my solution for this problem is to offer a seminar back-to-back after the InfoVis course, where interested students can choose a specific topic and prepare a presentation on it. This allows for more deeper discussions. Additionally, such a seminar is a good starting point for subsequent thesis work.

Typically, my courses terminate with an oral examination. Regarding our survey results, this is consistent with the examination practice of many colleagues coming from Europe, cp. Section 2.4.

4.3   Jason Dykes, City University London, UK

Introduction and Context: My background is in geography and the geo-sciences, where there is a strong tradition in the use of maps and graphics. Traditionally undergraduate courses in geography have taught and assessed manual skills in cartography and mapping activities and projects remain at the core of many geology degrees. Increasingly these activities are being augmented or replaced by learning that involves Geographic Information Systems (GISystems) and other digital techniques, but a focus on cartographic principles for designing effective maps and communicating geographic information effectively is still regarded as important.

My 'Visualization' Module. I have been teaching a Masters level module in 'Geo-Visualization' for eight years. The postgraduate module was originally developed to provide Geographic Information Science (GIScience) students with skills in creating, evaluating and using maps and graphics in their analysis and communication of geographic information. It replaced a cartography module and updated this with recent advances in the use of dynamic and interactive graphics for exploratory analysis.

During this time the GIScience community and GIScience students have become more aware of Information Visualization. Equally students from non GI programs have wanted to learn about cartography and visualization. It is challenging to develop a coherent module that is relevant to this range of students whilst embracing Information Visualization and a traditional cartography syllabus.

My module consists of 12 sessions with 3 hours contact time and 7 hours guided individual study. Students are expected to spend another 30 hours participating in assessed activities. The module is thus designed such that an average student spends 150 hours studying for 15 credits—-it uses a shared credit framework that does not conform to the requirements of the Bologna process [5]. The module is available to distance learners through a managed learning environment and digital resources ensure distance learners have an equivalent experience those who attend campus.

Many students go on to do research projects in visualization involving data sets and applied problems that feed my research and are subsequently used in teaching. This effective feedback loop is important in perpetuating bi-directional links between research and teaching.

Teaching Issues in Information Visualization at Dagstuhl: I don't know the content or culture of InfoVis or Computer Science education. This is one of the difficulties associated with having interests that lie between rapidly developing disciplines. But the Dagstuhl discussions and survey suggest that many of those involved are facing similar issues with which I am familiar. The survey draws attention to a range of approaches, topics, academic disciplines, text books involved.

There is great variation in the approaches and syllabuses of the courses reported in the survey. 'Information Visualization' is the most popular title, but course names that begin: 'Models and Algorithms for ... '; 'Data Visualization'; 'Computer Graphics and ... '; 'GeoVisualization' are evidently situated in different ways and will require different emphases. This is evident in Figure 1—a tag cloud showing relative occurrences of topics listed in courses included in the Dagstuhl survey.

Despite this texture, some consistent themes emerge from the survey and particularly the discussion at Dagstuhl. Three particular themes resonated with me and related to my experience. They are also evident in a report produced following an open workshop at IEEE Visualization 2006 in which I graphically depict the relationships between related courses with an Information Visualization emphasis [56]:

1. a trend/desire for evaluative and critique-based learning that can draw from a range of related disciplines;
2. an emphasis on 'learning through doing' as opposed to a transmissive approach to learning;
3. some concern about developing appropriate exercises and assessment for visualization classes.

I'll briefly consider these key issues and reflect upon their relationship with my teaching. Doing so may provide some synergies in the cross disciplinary educational mash-up that is developing in Information Visualization.

Critiquing: Critiquing involves students applying and developing their knowledge through evaluation and review. Robert Kosara argued convincingly for a critique-based approach to Information Visualization at Dagstuhl. He describes this as a 'highly interactive and human-centered way of designing things' [40]. An approach that evaluates graph drawing algorithms and evaluation criteria is documented in the Dagstuhl survey and described in Section 2.3 above.

As a learning device, critiquing may involve evaluating existing work, software or graphics, but importantly should involve the practical application of theory. I ask students to critique existing graphics, those that they have developed and software systems using a number of criteria, such as:

- Graphical integrity, graphical excellence and Tufte's 'theory of data graphics' [69]
- Using appropriate symbolism [6, 10]
- Map symbolism [45, 60]

- Map design [7]
- Use of colour [22]
- Interactivity and functionality [11, 41]
- Animation [21, 55]

Each of these sources provides useful criteria to structure critiques and against which judgments can be made. Amar and Stasko's framework [2] also offers opportunities and Dagstuhl has drawn attention to the scope for using more knowledge from InfoVis in developing critiquing criteria.

Paper Summaries. Presenting summaries of research papers is clearly a core activity in InfoVis teaching and learning. This fits the critiquing trend if the work is evaluative and assessed against existing theory and criteria. Tamara Munzner's generic 'what makes a good InfoVis paper' criteria as presented and discussed at the Dagstuhl seminar [52] could prove useful in this kind of learning activity.

The validity of critiquing depends upon the level of learning being supported. In the UK some of the more advanced levels of knowledge, such as those developed through critiquing, relate to the more advanced levels of learning. Masters level and level-3 courses require students to evaluate and so the approach is particularly appropriate. It can be used to provide feedback and for assessment. I've found that extended abstracts work well as a focus for quick and informal 'in class' critiquing that breaks up a learning session.

Learning by Doing:

"I hear and I forget, I see and I remember, I do and I understand" (Chinese proverb).

Participants in an Information Visualization seminar are likely to be persuaded that seeing is a powerful learning device. But 'learning by doing' may be even more effective. It is a form of active learning that is popular in education and there is particular opportunity for using this approach in visualization education. Doing can involve summarizing information, developing graphics or software or analyzing data sets and the Dagstuhl participants provide plenty of examples of these activities in their courses as we have seen in this paper.

Fieldwork is used to provide opportunities for learning by doing in the geosciences and has an important role in education [31]. There are parallels between the kind of observation, interaction and exploration that occur in fieldwork and those associated with the kind of exploratory analysis that visualization supports. I have explored some of these parallels and opportunities when developing learner-focused activities that emphasize learning by doing. The constructivist approach emphasizes learning through an interpretive, non-linear and recursive process in which active learners interact with their surroundings and the resources that are provided to help [19]. Such methods are regularly used in geoscience fieldwork and may result in 'deep understanding' of the kind experienced when researchers use visualization to interact with complex structured datasets. They seem particularly suited to visualization education.

Multiple perspectives are important in constructing knowledge and so group work is often employed in such activities. In this context, symbolic and graphical representations of knowledge and ideas and can be important in negotiating, mediating and constructing shared meanings [19]. This is particularly so if the graphics can be manipulated in an exploratory context as knowledge is derived. Consequently strong arguments exist for using software that is interactive and exploratory to develop opportunities for learner-centered constructivist activities [74].

I have developed exploratory visualization software for use in fieldwork in this context. It encourages 'learning by doing' through a series of linked cartographic, statistical and photographic views of a study area and a highly interactive interface through which these can be manipulated. It forms part of a learner constructed activity - the software is one of a series of resources made available to students who are expected to develop an approach to a problem relating to land cover and land use. The contrasts with passive or transmissive education are emphasized by the exploratory, student-led visualization and the ability to add data recorded in the field into the software for analysis [46]. Our evaluations show that the software and the constructivist activity that it supports are an effective learning device [47].

The Dagstuhl survey suggests that a number of similar activities may be taking place. It certainly seems that there is scope for using exploratory graphical software to support constructivist learning methods in visualization that give students the opportunity to learn actively.

A Portfolio Approach to Assessment: The issue of exercises and assessment was discussed at Dagstuhl primarily because some teachers were finding it difficult to set and assess meaningful exercises. Critiquing and 'learning by doing' provide scope for effective objective assessment that encourage active learning. My experience with portfolio-based assessment has been very positive in this context and compliments these methods well.

Portfolios involve students developing and collating annotated evidence of their capabilities throughout a course. They receive formative feedback on their work and use selected work to demonstrate that they have achieved a set of learning outcomes. Portfolio-based assessment can spread workload for staff and students, offers the opportunity for developmental feedback and review, and provides a synoptic view of what has been learned. It also results in a tangible end-product that students can show to colleagues and prospective employers— potentially—bringing learning to life and engendering pride.

I use portfolio-based assessment in the GeoVisualization module. Students participate in a practical exercise associated with each learning session that requires them to 'learn by doing'. Their work is discussed and improved. Selected work is submitted for formal formative feedback. At the end of term students are asked to submit a portfolio of three exercises and a reflective essay that uses these as items of evidence to demonstrate that module learning outcomes have been achieved. Amongst other competencies, outcomes require that students are able to ...

- explain the complex issues associated with GeoVisualization with clarity and from an informed perspective by drawing upon recent academic research;
- design maps and data graphics that are effective, informative and consistent and that exhibit graphical excellence and graphical integrity;
- use data graphics, maps and visualization tools to present and explore multifaceted data sets in a manner that is professional, informed and ethically sound;
- evaluate data graphics, maps and visualization tools by drawing upon principles and theories of design.

The approach seems to work nicely and addresses some of the issues discussed at the Dagstuhl meeting. It may be useful for those wishing to help students learn to critique and assess their developing skills. I have received favorable feedback from students and internal and external evaluators.

It seems particularly appropriate for developing skills in graphicacy where Tufte's concept of 'redesign' is a key element. Portfolios or long-term developing group projects that provide opportunities for feedback and critique can be very beneficial here. It should be noted that portfolios are frequently used in the arts where critiquing and redesign are key learning activities.

Conclusion: These arguments are personal perspectives, developed through reflections on discussions at Vis 2006 and the Dagstuhl seminar and the Dagstuhl survey, in the light of my teaching experience. I suggest a focus in Information Visualization education not on curriculum (which may vary to suit particular disciplines and student groups) but on general qualities and competencies that can be applied across a range of curricula. I've identified what some of these might be and presented some ideas on how they might be supported and assessed through the example of my GeoVisualization module.

The critiquing and 'doing' themes support the notion that underlying skills in the use and evaluation of graphics are broadly (I hesitate to say 'universally') valuable. They may be a way of helping Information Visualization educators bridge the multiple multi-disciplinary divides and certainly help justify the approach taken on my GeoVisualization module.

They may also help us deal with change—we are in a discipline where the specifics of curricula change very rapidly. Perhaps we're moving away from a text-book based model of teaching as disciplines change so quickly and student's expectations and abilities to access information increase. This is good news from the perspectives of information sharing and efforts to link research and teaching, which many are evidently doing very effectively, cf. Section 4.1. The bad news is that we need more flexibility in terms of course content and perhaps structure. This is no bad thing in itself, but is difficult to achieve when the levels of documentation that are expected by students and (in the UK at any rate) required for quality assurance and to gain approval for changes in provision are considered. Guided approaches in which links to a selected set of papers and examples are used to support learning against broad aims and outcomes provide a way forward. It is well worth using the URLs listed in the Dagstuhl survey to learn from colleagues who use this approach (see Stasko, Munzner and Heer's courses for example). The kinds of repositories of examples and teaching materials discussed and suggested in Section 4.1 will help, as will a focus on generic methods of teaching such as those that involve critique and active learning rather than developing monolithic curricula that will age rapidly in response to new developments. Portfolio-based assessment that involves the focused combination of a series of activities supports this flexible approach.

Adoption of the ideas discussed here would continue to move visualization education away from core Computer Science. Doing so will continue the trend of enabling more students to participate in visualization education and help address the difficulties associated with multi-disciplinary domains - how do we focus simultaneously on the concepts listed in our tag cloud of the scope of Information Visualization education (Figure 1)—computer science and algorithms, the science of perception and cognitive studies and concepts derived from the arts such as composition and design? The Dagstuhl survey has certainly helped inform my approach to visualization education. Perhaps this discussion will help the community when considering the nature of Information Visualization education and how to best it might be supported and developed. I'd certainly be delighted to debate the ideas and their relevance further.

5   Conclusion

This paper describes the results of our teaching survey based on the information given by the Dagstuhl attendees. It covers several aspects of offered InfoVis courses that range from different kinds of study materials to practical exercises. We have reproduced the discussion during the Dagstuhl Seminar and added our own experiences. In this regard, we have found that teaching InfoVis is challenging because it is a new and growing field. There exist a lot of open questions regarding the syllabus, a consistent theory, or the abstraction level of single topics. In consequence, it is also a great subject for teachers, not only for students: we are convinced that teaching InfoVis also leads to a better reflection on the topics and to new ideas which can induce new projects. Finally, we hope that this paper can serve as an interesting and helpful source for current and future InfoVis teachers.

Acknowledgments. We would like to thank all participants at the seminar [13, 36] for filling out the teaching survey and for the lively discussions. The survey was developed by Keith Andrews, and a first analysis of the results was also made by him. We thank him for his ideas and efforts.

References

1. Amar, R., Eagan, J., Stasko, J.: Low-level components of analytic activity in information visualization. In: Proceedings of the 2005 IEEE Symposium on Information Visualization - INFOVIS '05, October 2005, pp. 111–117 (2005)

2. Amar, R., Stasko, J.: A knowledge task-based framework for design and evaluation of information visualizations. In: IEEE Symposium on Information Visualization (InfoVis), Austin, TX, pp. 143–150. IEEE Computer Society Press, Los Alamitos (2004)
3. Amar, R.A., Stasko, J.T.: Knowledge precepts for design and evaluation of information visualizations. IEEE Transactions on Visualization and Computer Graphics 11(4), 432–442 (2005)
4. Andrews, K.: Course: Information Visualisation (2007), http://courses.iicm.tugraz.at/ivis
5. BENELUX Bologna Secretariat: About the Bologna Process (2008), http://www.ond.vlaanderen.be/hogeronderwijs/bologna/about/
6. Bertin, J.: Semiology of Graphics: Diagrams, Networks, Maps. In: Translation of Semilologie Graphique, University of Wisconsin Press, Madison (1983)
7. Brewer, C.A.: Designing Better Maps: A Guide for GIS Users. ESRI Press, Redlands (2005)
8. Card, S., Mackinlay, J., Shneiderman, B. (eds.): Readings in Information Visualization: Using Vision to Think. Morgan Kaufmann, San Francisco (1999)
9. Chen, C.: Information Visualization: Beyond the Horizon. Springer, Heidelberg (2006)
10. Cleveland, W.S., McGill, R.: Graphical perception: Theory, experimentation and application to the development of graphical methods. Journal of the American Statistical Association 79, 531–554 (1984)
11. Crampton, J.: Interactivity types in geographic visualization. Cartography and Geographic Information Science 29(2), 85–98 (2002)
12. Csallner, C., Handte, M., Lehmann, O., Stasko, J.: FundExplorer: Supporting the diversification of mutual fund portfolios using Context Treemaps. In: Proceedings of the 2003 IEEE Symposium on Information Visualization - INFOVIS 2003, October 2003, pp. 203–208 (2003)
13. Dagstuhl: Seminar 07221 Information Visualization – Human-Centered Issues and Perspectives (2007), http://www.dagstuhl.de/07221
14. Diehl, S., Görg, C., Kerren, A.: Preserving the Mental Map using Foresighted Layout. In: Proceedings of Joint Eurographics – IEEE TCVG Symposium on Visualization (VisSym '01), Eurographics, pp. 175–184. Springer, Heidelberg (2001)
15. Dykes, J., MacEachren, A.M., Kraak, M.-J.: Exploring Geovisualization. Pergamon Press, Oxford (2005)
16. Fayyad, U., Grinstein, G., Wierse, A.: Information Visualization in Data Mining and Knowledge Discovery. Morgan Kaufmann, San Francisco (2001)
17. Fekete, J.-D., van Wijk, J.J., Stasko, J.T., North, C.: The Value of Information Visualization. In: Kerren, A., Stasko, J.T., Fekete, J.-D., North, C.J. (eds.) Information Visualization. LNCS, vol. 4950, Springer, Heidelberg (2008)
18. Few, S.: Show Me The Numbers. Analytics Press (2004)
19. Fosnot, C.T., Perry, R.S.: Constructivism: A psychological theory of learning. In: Fosnot, C.T. (ed.) Constructivism: Theory, Perspective and Practice, 2nd edn., pp. 8–38. Teacher's College Press, New York (2005)
20. Grinstein, G., O'Connell, T., Laskowski, S., Plaisant, C., Scholta, J., Whiting, M.: Vast 2006 contest - a tale of alderwood. In: Proceedings of the 2006 IEEE Symposium on Visual Analytics, Science and Technology - VAST 2006, October 2006, pp. 215–216 (2006)
21. Harrower, M.A.: Tips for designing effective animated maps. Cartographic Perspectives 44, 63–65 (2003)

22. Harrower, M.A., Brewer, C.A.: Colorbrewer.org: An online tool for selecting color schemes for maps. The Cartographic Journal 40(1), 27–37 (2003)
23. Hauser, H.: Course: Information Visualization (2007), http://www.cg.tuwien.ac.at/courses/InfoVis/
24. HCC: HCC Education Digital Library (2007), http://hcc.cc.gatech.edu/
25. Heer, J.: Course: Visualization (2006), http://vis.berkeley.edu/courses/cs294-10-sp06/
26. ILOG Visualization Suite. ILOG, Inc. http://www.ilog.com/products/visualization/index.cfm, 2007.
27. InfoZoom: humanIT Software GmbH (2007), http://www.infozoom.com/enu/
28. Jankun-Kelly, T.: Course: Information Visualization (2006), http://www.cse.msstate.edu/~tjk/teaching/cse8990/
29. Keim, D.: Course: Information Visualization (2007), http://infovis.uni-konstanz.de/index.php?region=teach&event=ss07&course=infovis
30. Keller, P.R., Keller, M.M.: Visual Cues – Practical Data Visualization. IEEE Computer Society Press, Los Alamitos (1993)
31. Kent, M., Gilbertson, D.D., Hunt, C.O.: Fieldwork in geography teaching: a critical review of the literature and approaches. Journal of Geography in Higher Education 21(3), 313–332 (1997)
32. Kerren, A.: Generation as Method for Explorative Learning in Computer Science Education. In: Proceedings of the 9th Annual Conference on Innovation and Technology in Computer Science Education (ITiCSE '04), Leeds, UK, pp. 77–81. ACM Press, New York (2004)
33. Kerren, A.: Learning by Generation in Computer Science Education. Journal of Computer Science and Technology (JCS&T) 4(2), 84–90 (2004)
34. Kerren, A.: Course: Information Visualization (2007), http://w3.msi.vxu.se/~kerren/courses/lecture/ws06/infovis/
35. Kerren, A., Ebert, A., Meyer, J. (eds.): Human-Centered Visualization Environments. LNCS, vol. 4417. Springer, Heidelberg (2007)
36. Kerren, A., Stasko, J.T., Fekete, J.-D., North, C.: Workshop Report: Information Visualization – Human-centered Issues in Visual Representation, Interaction, and Evaluation. Information Visualization 6(3), 189–196 (2007)
37. Kerren, A., Stasko, J.T., Fekete, J.-D., North, C.J. (eds.): Information Visualization. LNCS, vol. 4950. Springer, Heidelberg (2008)
38. KidPad: University of Maryland (2007), http://www.kidpad.org
39. Kosara, R.: Course: Visual Communication in Computer Graphics and Art (2007), http://eagereyes.org/VisComm
40. Kosara, R.: Visualization Criticism: One Building Block for a Theory of Visualization. In: Kerren, A., Stasko, J.T., Fekete, J.-D., North, C. (eds.) Abstracts Collection – Information Visualization - Human-Centered Issues in Visual Representation, Interaction, and Evaluation. Dagstuhl Seminar Proceedings, Dagstuhl, Germany, vol. 07221, Internationales Begegnungs- und Forschungszentrum für Informatik (IBFI) (2007), http://drops.dagstuhl.de/opus/volltexte/2007/1136
41. Krygier, J.B., Reeves, C., DiBiase, D.W., Cupp, J.: Design, implementation and evaluation of multimedia resources for geography and earth science education. Journal of Geography in Higher Education 21(1), 17–39 (1997)
42. Lichtenbelt, B., Crane, R., Naqvi, S.: Introduction to Volume Rendering. Prentice-Hall, Englewood Cliffs (1998)
43. Ma, K.-L.: Course: Information Visualization (2006), http://www.cs.ucdavis.edu/~ma/ECS272/

44. Ma, K.-L.: Course: Information Interfaces (2007),
    http://www.cs.ucdavis.edu/~ma/ECS163/
45. MacEachren, A., Taylor, D.: Visualization in Modern Cartography. Pergamon,
    Oxford (1994)
46. Marsh, S., Dykes, J.: Using usability to evaluate geovisualization for learning
    and teaching. In: Proceedings of the GIS Research UK 13th Annual Conference
    GISRUK 2005, Glasgow (2005)
47. Marsh, S., Dykes, J.: Visualization software works as a focus for fieldwork! evalu-
    ating geovisualization software in a student-centred exercise with usability engi-
    neering techniques. Journal of Geography in Higher Education (submitted)
48. Miksch, S.: Course: Information Visualization (2006),
    http://www.ifs.tuwien.ac.at/~silvia/wien/vu-infovis/
49. Monmonier, M.: How to Lie with Maps, 2nd edn. University of Chicago Press,
    Chicago (1996)
50. Mullet, K., Sano, D.: Designing Visual Interfaces: Communication Oriented Tech-
    niques. Prentice-Hall PTR, Englewood Cliffs (1994)
51. Munzner, T.: Course: Information Visualization (2006),
    http://www.cs.ubc.ca/~tmm/courses/infovis/
52. Munzner, T.: Process and Pitfalls in Writing Information Visualization Research
    Papers. In: Kerren, A., Stasko, J.T., Fekete, J.-D., North, C.J. (eds.) Information
    Visualization. LNCS, vol. 4950, Springer, Heidelberg (2008)
53. North, C.: Course: Information Visualization (2007),
    http://infovis.cs.vt.edu/cs5764/
54. Plaisant, C., Fekete, J.-D., Grinstein, G.: Promoting insight-based evaluation of
    visualizations: From contest to benchmark repository. IEEE Transactions on Vi-
    sualization and Computer Graphics 14(1), 120–134 (2008)
55. Rana, S., Dykes, J.: Framework for augmenting the visualization of dynamic raster
    surfaces. In: EuroSDR Commission 5 Workshop on Rendering and Visualisation
    (Electronic Proceedings), Enschede, The Netherlands (2003)
56. Rushmeier, H., Dykes, J., Dill, J., Yoon, P.: Revisiting the need for formal edu-
    cation in visualization. IEEE Computer Graphics and Applications 27(6), 12–16
    (2007)
57. Schroeder, W., Martin, K., Lorensen, B.: The Visualization Toolkit – An Object-
    Oriented Approach To 3D Graphics, 4th edn. Kitware, Inc (2006)
58. Schumann, H., Müller, W.: Visualisierung – Grundlagen und allgemeine Metho-
    den. Springer, Heidelberg (2000)
59. Shakshuki, E., Kerren, A., Müldner, T.: Web-based Structured Hypermedia Al-
    gorithm Explanation System. International Journal of Web Information Sys-
    tems 3(3), 179–197 (2007)
60. Slocum, T.A., McMaster, R.B., Kessler, F.C.: Thematic Cartography and Geo-
    graphic Visualization, 2nd edn. Prentice-Hall, Englewood Cliffs (2004)
61. Spence, R.: Information Visualization, 1st edn. ACM Press, New York (2001)
62. Spence, R.: Information Visualization: Design for Interaction, 2nd edn. Prentice-
    Hall, Englewood Cliffs (2007)
63. Spotfire: TIBCO Software, Inc. (2007), http://spotfire.tibco.com
64. Stasko, J.T.: Course: Information Visualization (2006),
    http://www.cc.gatech.edu/~stasko/7450/
65. Tableau Desktop: Tableau Software, Inc. (2007),
    http://www.tableausoftware.com
66. TableLens: Inxight Software, Inc. (2007),
    http://www.inxight.com/products/sdks/tl/

67. Tufte, E.: Envisioning Information. Graphics Press, Cheshire (1990)
68. Tufte, E.: Visual Explanations: Images and Quantities, Evidence and Narrative. Graphics Press, Cheshire (1997)
69. Tufte, E.: The Visual Display of Quantitative Information, 2nd edn. Graphics Press, Cheshire (2001)
70. Ward, M.: Course: Data Visualization (2006),
    `http://web.cs.wpi.edu/~matt/courses/cs525D/`
71. Ware, C.: Information Visualization: Perception for Design, 2nd edn. Morgan Kaufmann, San Francisco (2004)
72. Yang, K.: Course: Information Visualization (2007),
    `http://www.cs.uncc.edu/~jyang13/infovis2007.html`
73. Yi, J.S., Kang, Y.a., Stasko, J., Jacko, J.: Toward a deeper understanding of the role of interaction in information visualization (Paper presented at InfoVis '07). IEEE Transactions on Visualization and Computer Graphics 13(6), 1224–1231 (2007)
74. Zerger, A., Bishop, I.D., Escober, F., Hunter, G.: A self-learning multimedia approach for enriching gis education. Journal of Geography in Higher Education 26(1), 67–80 (2002)

Creation and Collaboration: Engaging New Audiences for Information Visualization

Jeffrey Heer[1], Frank van Ham[2], Sheelagh Carpendale[3], Chris Weaver[4], and Petra Isenberg[3]

[1] Electrical Engineering and Computer Sciences,
University of California, Berkeley,
360 Hearst Memorial Mining Building, Berkeley, CA 94720-1776, USA,
`jheer@cs.berkeley.edu`
[2] IBM Research, Visual Communications Lab,
1 Rogers Street, Cambridge, MA 02142, USA,
`fvanham@us.ibm.com`
[3] Department of Computer Science, University of Calgary,
2500 University Dr. NW, Calgary, AB, Canada T2N 1N4,
{`sheelagh, petra.isenberg`}`@ucalgary.ca`
[4] GeoVISTA Center and the North-East Visualization and Analytics Center,
Department of Geography, Penn State University,
302 Walker Building, University Park, PA 16802, USA,
`cew15@psu.edu`

1 Introduction

In recent years we have seen information visualization technology move from an advanced research topic to mainstream adoption in both commercial and personal use. This move is in part due to many businesses recognizing the need for more effective tools for extracting knowledge from the data warehouses they are gathering. Increased mainstream interest is also a result of more exposure to advanced interfaces in contemporary online media. The adoption of information visualization technologies by lay users – as opposed to the traditional information visualization audience of scientists and analysts – has important implications for visualization research, design and development. Since we cannot expect each of these lay users to design their own visualizations, we have to provide them tools that make it easy to create and deploy visualizations of their datasets.

Concurrent with this trend, collaborative technologies are garnering increased attention. The wide adoption of the Internet allows people to communicate across space and time, and social software has attained a prominent position in contemporary thinking about the Web. For example, one can think of software teams distributed over different time zones or multiple people collaborating to build an online encyclopedia. Furthermore, collaborative issues are not limited to the web: novel display and interaction technologies, including wall-sized and tabletop interfaces, introduce new possibilities and challenges for co-located collaborators. An increased need for specialization means that we can no longer rely on a single person to perform deep analyses of complex phenomena. These

A. Kerren et al. (Eds.): Information Visualization, LNCS 4950, pp. 92–133, 2008.

developments signify an increased desire for collaboration around complex data, yet, information visualization tools are still primarily designed according to a single user model. To meet the demands of an increasingly diverse audience, the design of information visualization technologies will have to incorporate features for sharing and collaboration.

In this paper, we discuss creation and collaboration tools for interactive visualization. Our goal is to begin to characterize the increasingly diverse audience for visualization technology and map out the design space for new creative and collaborative tools to support these users. In section 2 we classify the expanding user base for visualization technologies by looking at their skills, goals and the data they are trying to analyze. We then take a look at existing information visualization tools and classify them along these dimensions. In sections 3 and 4 we examine the new collaborative trends. Section 3 discusses co-located collaboration, while section 4 explores the area of distributed, asynchronous collaboration on the Web. Finally, we conclude by considering the ways the research community should respond to these developments.

2 End-User Creation of Visualizations

The term "end-user visualization" encompasses a broad range of visualization users and use-cases. For example, a marketing executive might create an overview of the sales in different product segments to show to his manager, a scientist may create a coordinated visualization application to study a biomedical dataset, or a Facebook user may present her social network in a visualization on the site. All these use cases involve different types of users employing information visualization to tackle different types of problems. If we wish to provide end-users with the ability to construct and deploy custom information visualizations of their own data, we need an understanding of these users, their goals, and their data. In the following sections, we will broadly classify each of these dimensions. Note that we do not intend to construct a formal taxonomy of users. Instead, our goal is to broaden the discussion on who our users are and how visualization can help them.

2.1 Data

Scientific, geographic, economic, demographic, and other domains of human knowledge produce vast amounts of wildly different forms of information, varied in terms of both individual interest and broad social importance. Visualization seeks to provide perceptually and cognitively effective tools to display and interact with these different kinds of data. Data is commonly categorized by inherent complexity (e. g., data homogeneity, number of dimensions) or size. In this section, however, we consider data from the perspective of users by categorizing three different kinds of data in terms of potential audience.

Personal Data: Personal data encompass all types of organized information collections that are of personal interest to a particular user, but less interesting to a broader community. This may involve data on user-owned media (such as DVD collections or playlists), data on life organization (financial data or address books) or data related to hobbies and general interests (photo collections, fitness schedules or coin collections). Visualizing personal data might not always lead to deep new insights about the data itself. In such cases, the visualization instead may serve more as a compact visual artifact that can be used to remember certain events in ones life and serve as a visual representation of self [95]. These visual representations of self may then be used as online avatars, or simply as catalysts for storytelling, much like photo albums.

Community Data: By community data we mean data that might be relevant to a broad community of users due to similar interests or general appeal. Examples of community data include the content of political speeches, the number of users online in a World of Warcraft realm, or voting results per county. Often this type of data has a social component associated with it: data might be related to a social application such as Facebook or MySpace [45], contain statistics on a large population as with census data [43], or may be related to current events [104]. Precisely because community data has a lot of general appeal it will often generate a lot of discussion.

Scientific Data: Scientific data is data that is of interest to a (relatively) small number of specialists. Traditionally, information visualization has focused on the sciences, because they generate a wealth of structured and often numerical data in ready need of analysis. This makes them very suitable to mathematical analysis techniques and visual mapping. In the humanities, however, most information comes in unstructured raw text format. If we want visualization to be applied in domains such as literature and political science, we will need to define suitable pre-processing techniques that can extract meaningful information from a body of text. This will often require some amount of natural language processing or expert input. While there are a few applications of information visualization to data from the humanities (e. g., [101]), the area remains largely untapped despite substantial promise to yield many useful techniques with applicability to many different areas of everyday life.

Interplay of Data Types: Note that the distinction between types of data is not always clear cut and many data sets could fall into different categories depending on their use. For example, a community data set on World of Warcraft users and their interactions might be considered a scientific data set by social scientists, while the personal data of celebrities might have a broad general appeal. Visualizations of all these types of data can be shared, albeit for different purposes. Personal data might be shared with other users as a means of personal expression. Community data is often shared to spark broad discussion, while scientific data often needs to be shared because it is too complex for one person to analyze on their own or because it requires multiple specialized skills to analyze.

The recent trend toward visual analytics [91] is driven by the increasing need to support open-ended management and exploration of large, loosely-connected, and often unstructured information sources as well as the smaller, isolated, structured data sets typical of information visualization applications. Information collection often involves assembling "shoeboxes" of loosely related nuggets and data sets [107]. Visual analysis of information occurs by following chains of evidence, evaluating formal hypotheses [27], testing competing explanations [86], or telling stories [37] using visual metaphors to convey relationships and dynamics. These activities are particularly challenging in intelligence analysis, emergency management, epidemiology, and other critical areas that involve high-dimensional abstract information [83] and large geospatial datastores [36]. However, the heterogeneous and idiosyncratic nature of the data sets and analysis activities in these endeavors are similar to those in everyday domains, making it likely that the outcomes of visual analytics research will translate readily into visualization approaches that will help to engage broad audiences.

2.2   Skills

Novice Users: By novice users we mean users who have experience operating a computer, but no experience with programming in general, let alone programming visualization techniques. The vast majority of novice visualization users act as consumers: they will interact with the visualization within the possibilities offered but will rarely extend existing functionality to suit their analysis needs. If we want these users to be able to produce visualizations, we have to take care to make this process as easy as possible. Some points of consideration when designing visualizations for novice users are:

Data Input: We cannot expect a novice user to write their own data parser, write database queries that export data to a particular format or understand the file formats for more complex data types. Most novice users seem to take to using spreadsheet programs such as Microsoft Excel to store and analyze their data. One useful input format then, is a simple tab delimited input file, as this format is both human readable and can be directly copied from the spreadsheet editor.

Automatic Selection of Visualization Type: Novice users have no experience designing visual mappings and may even choose mappings that produce nonsensical visualizations. Recurring examples include the use of line charts over categorical data dimensions, for which a bar chart would be a better choice, and using a pie chart for data that do not form part of a whole. For this reason, visualization techniques geared towards novice users should at least partly automate the selection of visual metaphors. This may involve analyzing the data dimensions to see if there are any ordinal attributes, check for aggregated variables and totals, and examine values in dimensions for possible hierarchical structure [59,60].

Useful Defaults: Novice users likely will not spend time tuning an ugly looking visualization to fit their needs. It is therefore important to provide a set of sensible defaults for data and view parameters (such as scales, colors, item sizes and viewpoints) to help constrain the parameter space that users have to explore. Multiple combinations of these parameters can be offered by providing a preset list. As an added bonus, a good set of presets can show users what is possible and educate them on what is sensible.

Contextual Information: With contextual information we mean visual items that explain to the user what data is being mapped to the screen and what encodings are being applied. This involves legends, scales, labels, pop-ups, titles and explanations of visual mappings. Although visual graphics in print media take great care to provide contextual information, interactive visualizations are often lacking in this respect because most of the design attention is focused on the visual mapping itself.

Savvy Users: By savvy users we mean people who have experience performing relatively sophisticated data organization and manipulation, using a combination of manual processing and limited amounts of programming or scripting. Because savvy users are a small but non-trivial part of the population of visualization consumers, they are a critical bridge between experts and novices. As such, savvy visualization users may act variously as:

- *experts* who train or guide novice users in the use of particular visualizations by clarifying exploratory and analytic functionality in terms of interface appearance and behavior,
- *designers* who plan, construct, debug, test, and deploy new visualizations for ongoing evaluation and routine operation by novice users,
- *end-users* who can bring more extensive experience to bear when using existing visualizations to analyze data from their own knowledge domains, to browse data with which they are less familiar, and to share their results with others, and
- *explorers* (or *user-designers*) who combine the roles of designer and end-user by extending and redesigning visualizations on the fly during open-ended exploration of their data.

Expert Users: By expert users we mean people who have extensive experience with interactive graphical software development and the theory and application of data modeling, data processing, and visual data representation. As such, visualization experts may act both as:

- *researchers* who invent, specify, and evaluate methods for accessing, querying, rendering, and interacting with data, often with an eye toward extending and enhancing the functionality of existing visualization systems and tools, and

– *developers* who design and implement visualization modules, toolkits, systems, and tools of various sizes and scopes, often adapting and integrating existing functionality from other visualization toolkits and systems.

In particular, visualization research frequently involves the development of prototypes for evaluating the correctness, flexibility, and performance of new data processing algorithms and the usability and utility of new interaction techniques.

Facilitating the interdependent needs of novice, savvy, and expert users is a key part of supporting broader audiences for information visualization. The number of people who can act as visualization designers or visualization developers – let alone the core visualization researchers who by necessity often fill these roles – is rapidly becoming overwhelmed by demand for visual tools brought on by blossoming public awareness of the power and accessibility of information visualization techniques. It will become increasingly necessary to provide users of all skill levels, including novices, with the capability to explore and analyze data sets of personal and professional interest without direct assistance from traditional visualization practitioners. However, understanding how to design accessible yet flexible software artifacts for individual visual exploration and analysis is only half of this equation. Social organization of visualization roles through collaboration and other means, as described later in this paper, is the critical second half.

2.3 Goals

One of the traditional rationales for information visualization is that the human visual system has high input bandwidth and has evolved as an excellent tool for spotting patterns and outliers in our surroundings. If we then map large amounts of data into visual form, we can use these innate human abilities to explore the data to find patterns that would have been exceedingly difficult to identify through purely automated techniques. A current prominent example is bioinformatics research that visually explores gigabytes of gene experiments to investigate the mechanisms that drive a particular disease. Such "explorative" use-cases have dominated most of the research in visualization over the past two decades. Explorative use can either be open-ended, where the user wants to browse their data without having a predefined question in mind, or analytically driven, in which the user has a particular question in mind and uses the visualization to answer it. Often times these two types of exploration will be intertwined: a user will explore a previously unknown data set without a particular question in mind, stumble on an interesting data point and then use the analytic features in the visualization to either answer the question or redirect their open-ended exploration.

Exploration and Analysis: Recent visualization environments have begun to offer users various degrees of interactive control over different parts of the entire information interface design process, thereby opening up possibilities for much deeper exploration of data. Such environments allow computer-savvy user-designers to interactively access data, create, layout, and coordinate views, and connect data to views. Design typically occurs directly within the interface that contains data views, and often take effect immediately without the need for a separate compilation or build stage. This live, amodal approach to interface design allows users to switch rapidly between building and browsing tasks during exploration and analysis. The result is a form of exploration that is free form and open-ended, particularly during initial inspection of newly encountered data sets.

IVEE [2], DEVise [58], DataSplash [71], Snap-Together Visualization [69], GeoVISTA Studio [88], Improvise [102], and Tableau/Show Me [60] are a few of many well-known visualization environments that support open-ended data exploration to various degrees. Such environments typically consist of a graphic user interface on top of a library of visualization components which may or may not be exposed as a visualization programming toolkit in its own right. This combination of user interface and underlying library can enable open-ended exploration in a very broad sense if it bridges the activities of visualization users performing various roles with different levels of expertise, whether as individuals or in collaborative groups.

To connect developers and designers, a key advantage for open-ended exploration is an extensible library that provides an application programming interface (API) for adding new software modules for various visualization components (including data access, queries and other data transformation algorithms, views, and visual data encodings). In particular, the most useful APIs support the definition of new data transformation operators—including appropriate input and output data object types—that give designers the ability to express rich relationships between data, queries, and views. This requirement is essential for applying newly discovered visualization techniques to emerging sources and forms of information, without needing to constantly architect and implement new toolkits (and retrain visualization designers in their use).

To connect designers with users, the user interface must support the ability to access data sets (and metadata) from local or remote sources in various formats, create and position views on the screen, specify how navigation and selection affects views, specify queries on data, parameterize queries in terms of interaction, and attach data sets and queries to views. In particular, designers should be able to specify the appearance and behavior of their visualizations directly within the user interface, without resorting to programming or other workarounds for interface limitations. To do otherwise would effectively require that designers be trained as developers.

User interfaces that truly support open-ended exploration would exceed the requirements of basic visualization design and operation by: supporting live building of complete browser interfaces, including immediate designing, debugging, and testing of intended functionality; facilitating collaboration between end-users and designers to turn analytical questions into structural changes (through remote, nearby, or side-by-side efforts to communicate and effect rapid visualization prototyping and polishing); and enabling rapid switching between building and browsing to perform more extensive exploratory visualization by modifying visualization views and queries on the fly. In particular, it is highly desirable for explorers to be able to see all raw data quickly to make decisions about how to visualize it, rapidly create and lay out views, rapidly attach data and queries to views, rapidly modify queries, store, copy, and reuse views, copy-and-paste/drag-and-drop visualization components, and use macros to build common multiple view constructions. Many of these capabilities are also desirable for non-exploring designers who prepare visualizations for domain analysts.

In all of this, availability of common and familiar interface functionality is essential to broad adoption. The user interface should run in the user's normal working environment, require no programming or design activities, and provide a way to disseminate analytical results. For communication and collaboration, it is highly desirable for the user interface to run easily on any platform, allow visualizations to be opened and saved as normal documents for sharing between users, and provide the ability to bookmark or screen capture visualizations in different graphical states.

Communication: At the other end of the spectrum of information visualization goals is the "communicative" use-case, where the main user goal is simply to convey a message to others. This use-case is already present in many traditional media: think of diagrams explaining the numbers behind a news story or a bar chart that has been included in a slideshow presentation. Although these particular representations are fairly static because of the affordances of the media used, this does not mean that communicative information visualization is limited to static visualizations. Interactivity is a very useful means of engaging users and may make them more receptive to a particular message. However, interactivity also poses some problems when communicating visualizations, because it's hard to reproduce interactive features in a static medium. Many information visualizations use tooltips and mouse-overs to provide contextual information, offer the user different viewpoints on the data, and allow for dynamic analysis of data. Videos alleviate this problem only in part, because it is often hard to follow what is going on and much of the context in the exploration process is missing. Simply sharing findings using static representations of interactive visualization is therefore not the optimal solution, and we would do better to consider these issues beforehand when designing visualizations for communicative use.

Apart from traditional mass communication, the communicative use-case also plays a pivotal role in collaborative applications, especially ones that are non-collocated and/or asynchronous. If the analysts do not share the same time or space it is important for them to be able to communicate findings and bring each other up to speed on the current state of the process quickly. Furthermore, each of these types of collaboration (collocated or distributed) has its own type of requirements, which we discuss in-depth in sections 3 and 4. In general, communicative use of information visualization usually involves a small investment of time on the user end, with a small but guaranteed payoff. On the other hand, explorative use involves a large amount of investment in tools, training and time, while the (potentially high) payoff is not always guaranteed. (See also [93] for a discussion of these tradeoffs.)

In the next section, we consider a number of representative tools that help us meet these differing goals of communication and exploration.

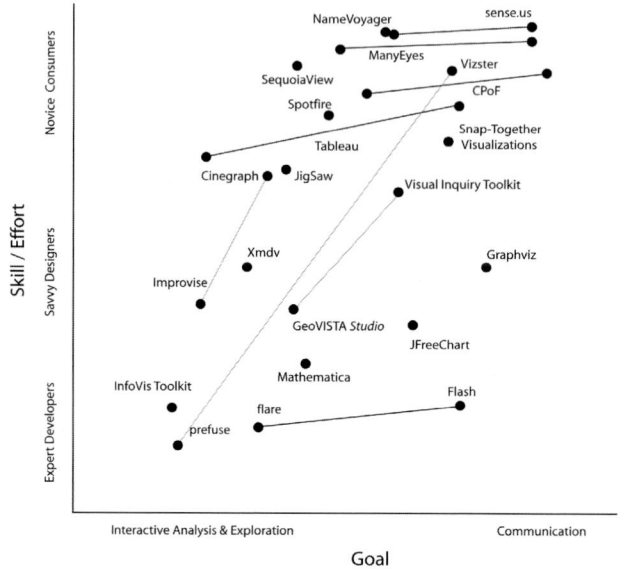

Fig. 1. A few of the many available information visualization tools, roughly mapped according to targeted end-user and targeted goal. Light lines connect toolkits and development environments to examples of visualizations created in them. Dark lines roughly capture similar ranges of user/goal targets for relevant tools.

2.4   Tools

Note that most real-world uses of information visualization will form a combination of the use-cases and roles described in the preceding sections. A researcher might program a new visualization technique to explore his complex data and then present findings to a manager by sending a screenshot. In this case the researcher takes on the roles of both consumer and developer and performs both exploration and communication. Most current information visualization tools and toolkits are geared towards one particular user skill and goal, although a recent trend towards more flexible tools can be observed. To illustrate the rough classification outlined in the previous subsections, in this section we give an indicative sample of an end-user visualization tool for each user skill and goal combination. Figure Fig. 1 illustrates a number of available visualization tools categorized according to the skill level of the target user base and the degree to which the tools support analytic and communicative tasks. Systems that span a range of tasks or skills are presented as line segments indicating the range of users and usage.

Expert Communication: In the bottom right corner of the matrix we find information visualization tools and toolkits that are geared towards communicative use, but assume a significant amount of knowledge on suitable visual techniques and their implementation. One such toolkit is Adobe's Flash development environment. Flash is an browser based interactive graphical design tool. Because of its ease of online deployment, it is particularly suited to communicating messages in a graphical way. In fact, many of the interactive graphics on the Internet today are Flash-based. However, Flash does not offer the developer tools that would support structured data exploration. (In fact, it offers only the most basic of data structures.) Moreover, its timeline-based design environment is not particularly well-suited to interactive visualization development.

This situation has improved with the release of Actionscript 3 and Flex, offering a more advanced programming model and a full-fledged user interface development package. The Flare toolkit [41] implements basic visualization capabilities within Flash, making it easier to develop interactive information visualizations, while still retaining the benefits of Flash, such as its relatively lightweight means of online deployment.

Savvy Communication: Information visualization tools in this category (middle right) allow users to create and share complex information visualizations, but require a base level understanding of computer programming and information visualization. A concrete example of such a tool is AT&T's GraphViz [33] library, which allows users to generate static images of graphs but requires some programming effort to integrate it with existing applications because it uses a proprietary data format.

Use-cases for GraphViz often involve reporting engines that need to be able to display networked data of some sort. Many of the features in GraphViz are geared towards presentation instead of exploration. For example, it is possible to heavily customize node rendering. Special care has been taken to avoid label overlapping, as this would make static images completely unreadable. Both of these issues are less of a problem in interactive systems in which users can use tooltips to get more information, or zoom into a dense cloud of labels to remove overlap.

Novice Communication: Until recently, if novice users wanted to share information visualizations with others they would be limited to taking screenshots of information graphics for sending by e-mail, etc. This mode is often sufficient if the goal is one-way information dissemination. For example, a pie chart may be included in a presentation, or an advanced information graphic may be printed in a newspaper. However, this mode of publication fails if users want to collaboratively analyze a complex data set.

Recent tools like Many Eyes [97], Sense.us [46], Swivel.com, and Spotfire Decision Site Posters make this process much easier, allowing easy sharing of interactive visualizations. As discussed in greater detail in section 4, users of these systems can share a particular state of a visualization encoded as a URL and add custom annotations and comments while still having access to the interactive features of the visualization. This makes it possible to quickly switch between analysis and communication, a necessity for successful collaboration.

Expert Exploration and Analysis: The bottom left corner of the matrix contains visualization software that supports deep and broad exploration of the space of visualization techniques, as well as more focused exploration and analysis of particular data sets. Such flexibility in the overall process of visualization almost always requires substantial expertise, typically requiring programming skills. As a result, visualization software for exploration by experts often takes the form of toolkits that are written in a popular programming language but that encapsulate well-known visualization components and techniques.

One such toolkit, prefuse [44], provides a Java-based library of visualization building blocks including views, visual encodings, processing algorithms, multi-view coordinations, and a common data model that supports tables, trees, and graphs. Graphs, hyperbolic trees, treemaps, and scatter plots support accessing, filtering, rendering, and displaying data using a variety of layout and distortion algorithms.

Similarly, the InfoVis Toolkit [34] is a set of Java visualization components designed around OpenGL and a data model that represents tables, trees, graphs, and metadata in column format for efficient selection, filtering, visual encoding, and coordination. Views include scatter plots, parallel coordinate plots, treemaps, and a variety of node-edge tree and graph displays that can incorporate fisheye lenses and dynamic labeling of items. Visualizations created in the toolkit display textboxes, sliders, and other controls alongside views for dynamic editing of visual encodings.

The extensible programming interfaces of both toolkits and those like them provide a means to incorporate new components and techniques, in essence expanding the scope of exploration, considered broadly, to include the results of future visualization research.

Savvy Exploration and Analysis: As described in the previous section, visualization in the expert exploration category revolves more around programming rather than around interaction in integrated user interfaces intended for designing and building tools. Research on integrated visualization environments focuses on packaging the exploratory capabilities of toolkits in ways that are accessible to users who are visualization savvy but not necessarily visualization experts.

For instance, Improvise [102] is a self-contained Java application that appears and behaves like other office productivity applications based on the multiple document desktop metaphor. Users build Improvise visualizations by interactively constructing the data, queries, views, and coordinations of tools that can be saved, opened, copied, and shared as self-contained Extensible Markup Language (XML) documents. Users browse visualizations using the mouse and keyboard to navigate and select data items in multiple coordinated views.

Similarly, GeoVISTA Studio [88] is an integrated visualization development environment for building geovisualizations interactively using a graph-based visual coordination editor. Any component that conforms to the JavaBeans specification can be a view. Development of new views by the community of GeoVISTA Studio users has resulting in a large library of views utilized in numerous visualizations. A particular strength of GeoVISTA Studio is its extensive functionality for representing and displaying geospatial information (based on the GeoTools [57] open source Java GIS toolkit).

The combination of browsing with rapid, iterative building in a single application (much like in spreadsheet programs) enables improvisational visualization, in which it is possible to design and evaluate different ways of analyzing particular data sets in the form of rapid prototypes having more concrete and stable collections of analytic functionality.

Novice Exploration and Analysis: As far as we know, there are no tools that truly allow novice users to interact with their data in the broadest sense of exploration. This may result from an apparent fundamental tradeoff between flexibility and accessibility in visual analysis, in that increased expressiveness necessitates greater expertise when it comes to data manipulation and visual representation. Even in savvy exploration-analysis tools like Improvise that strive for a balance between these factors, reproducing many common visual components and techniques currently requires a high degree of visual language expressiveness that necessitates a corresponding high level of expertise beyond that of most novice users. Conversely, novice analysis-communication tools like Many Eyes seek to increase visualization flexibility for broad audiences keenly interested in modest analytic expressiveness as a means to better communicate ideas about information. In between, analysis tools like Tableau/Show Me successfully occupy analysis niches that provide bounded but particularly useful forms of data interaction to relatively broad audiences who are sufficiently motivated to devote time and effort to modest training. It may well be that open-ended exploration tools for novices will evolve from future research into ways of combining these three seemingly complementary directions.

2.5   Directions

Current end user visualization tools are becoming more and more flexible in the types of scenarios and goals they can handle. Tools like Many Eyes allow novice users to create advanced visualizations with very little effort and also support communicative use-cases by allowing flexible sharing of visualization states. Tools like Improvise allow tight integration of many different types of visualizations, but require some programming skills on the side of the end-user, an expectation that is not always reasonable of domain experts dealing with the visualization. Tableau allows end users to set up and pivot different types of basic visualizations in a fairly intuitive manner and the recent addition of Tableau Server allows sharing of and commenting on these visualizations in an online environment, making it also suitable for communicative purposes. Although flexible, the only visualization types allowed are 2-dimensional small-multiple displays, which limits the visualization and analysis types to basic business graphics.

In our opinion, the ultimate goal of letting novice users flexibly specify their visualization needs and couple different types of views together has not been fully realized yet. We expect that users' visual literacy will increase as information visualization becomes more mainstream, and will start demanding advanced visualizations beyond the trusted bar chart. Integrating advanced visualizations in an flexible, collaborative and easy to understand framework for open-ended exploration and analysis is an important and solvable problem. We expect this solution will have important implications for many areas of human endeavor that necessitate the handling of complex data.

3   Co-located Collaborative Visualization

Given the choice, it is common and natural for people to work together. This is not a new phenomenon. Small groups of people gather for all kinds of reasons including many that are work related; such as to get a job done faster, to share expertise for a complex task, and to benefit from different insights from different people. Also, when one considers the rapid growth in size and complexity of datasets, it is not surprising that increasingly the practicality of an individual analyzing an entire data-set is becoming unrealistic. Instead, the expertise to analyze and make informed decisions about these information-rich datasets is often best accomplished by a team [91]. For instance, imagine a team of medical practitioners examining a patient's medical record to plan an operation, a team of biologists looking at test results to find causes for a disease, or a team of businessmen planning next year's budgets based on a large financial dataset. All of these situations involve a group of people making use of visual information to proceed with their work. Research towards supporting these team-based information processes will expand the situations in which information visualization can be used and is part of considering how to best support people in their normal everyday information work practices.

This section draws from a wide variety of literature to shed light on questions and issues that need to be considered during the development of co-located collaborative information visualizations. We do not consider this discussion to be exhaustive; rather it is our intention that the discussion will form the beginning of design guidelines and considerations that will be modified and extended through future research in collaborative information visualization.

Research in information visualization draws from the intellectual history of several traditions, including computer graphics, human-computer interaction, cognitive psychology, semiotics, graphic design, statistical graphics, cartography, and art [64]. The synthesis of relevant ideas from these fields is critical for the design and evaluation of information visualization in general and it is only sensible to think that fields concerned with collaborative work also add valuable information to our understand-ing of requirements for collaborative information visualization systems. Our sources include work in co-located collaboration in computer supported cooperative work [39,53,75,73,76,77,80,81,82,90], information visualization [85,105,109,110,111], and empirical work investigating collaborative visualization use [61,68,72].

The organization of this section is as follows. A brief overview of existing research that relates to co-located collaborative information analysis is given in section 3.1. Next, section 3.2 discusses the impact of recent advances in hardware configurations and section 3.3 focuses on more general human computer interaction issues important for the support of the co-located collaborative process, primarily drawing upon computer supported co-located collaborative literature. Then section 3.4 presents information visualization specific issues that may need re-consideration in light of co-located collaborative applications.

3.1   Related Research

Co-located collaborative information visualization is a relatively new and still under explored research area. Only a few tools designed specifically to support synchronous collaboration between co-located people using visualizations to explore information have emerged thus far. These are discussed below. However, as noted below, existing visualization tools designed from a single-user perspective have been studied with co-located collaborative tasks [61]. There has been considerable research in the area of scientific visualization in distributed systems (see [38] for an overview). Recently, there has been new primarily web-based research on asynchronous distributed collaborative information visualization systems. This new direction is the focus of section 4.

Co-located Collaborative Visualization: The Responsive Workbench was one of the first visualization systems for developed co-located collaboration around a large horizontal surface [103]. The responsive workbench is a virtual reality environment in which the displayed 3D scene is seen through shuttered glasses and interaction is achieved with a glove which has an attached Polhemus sensor on the back. Agrawala et al. [1] extended this workbench to support two simultaneous users. Several scientific visualization applications were developed for this platform including fluid dynamics and situational awareness applications.

On tabletop displays information visualization interaction techniques have been used to support co-located people in information sharing and exploration tasks. DTLens [35] provides a local non-linear magnification technique enabling multiple lenses for up to four people with for two-handed interaction. Personal Digital Historian uses radial layouts to display photos, video and text documents to supports conversation and story telling for small groups of people [84].

Studying Collaborative Use of Information Visualizations: While research on collaborative data analysis using information visualizations is relatively scarce, collaborative use of existing single user systems has been studied. Mark and Kobsa [61] conducted a user study in which they observed pairs working in co-located and distributed settings with two different visualization systems designed for single users. Their findings suggest that the benefit of collaborative vs. individual problem solving was heavily dependent on the visualization system used and also that, in general, groups were better than individuals working alone at locating errors. From this study, they derive a model for the collaborative problem-solving process. Their model consists of an iterative sequence of five stages: parsing a question, mapping variables to the program, finding the correct visualization, and two validation stages. From studying collaborative work on scientific visualizations in virtual environments using CAVEs, Park et al. [72] report a five-step activity model that was common for the observed collaboration sessions. Their study also noted that participants showed a strong tendency for independent work, if the option was available. Isenberg et al. studied co-located collaborative data analysis scenarios and posit an eight-process framework that relates to previous work on the Sensemaking Cycle [17] and the two studies by Mark and Kobsa [61] and Park et al. [72]. However, a common temporal order of analysis processes as posited by some previous work did not emerge.

3.2   Choosing Hardware to Support Co-located Collaboration

We start with a discussion of hardware because some of the recent interest in co-located collaboration is at least in part due to new hardware innovations.

Display Size: In information visualization, the size of the available display space has always been problematic for the representation of large datasets (e. g., [65]). In a common desktop environment, typically a single user will use all available screen space to display their visualization and, most commonly, this space will not be sufficient. Frequently, visualization software will include interactive features to help the user cope with limited display space. It seems sensible to think that, if we are going to adequately support collaborative or team exploration of visualizations, available display space will be an important issue. In collaborative systems, screen space not only has to be large enough for the required information display, it might also have to be viewed and shared by several users. As the number of people using a shared information display grows, the size of the display and workspace needs to be increased in order to provide a viewing and interaction area that gives adequate access to all group members.

Display Configuration: Several configuration possibilities exist that could increase the amount of available display space, all of which will affect the type of visualization systems possible and the type of collaboration work that would be most readily supported. Many types of configurations are possible; for instance, one could provide team members with interconnected individual displays, as in the ConnecTable system [89], or one could make use of large, interactive, single-display technology, like display walls or interactive tabletop displays (e. g., [90]). An additional possibility is to link wall, table, and personal displays (e. g., [105]), or to consider immersive displays (e. g., [72]). The type of setup most appropriate for an information visualization system will depend on the specific task and group setup. For example, individual interconnected displays allow for private views of at least parts of the data which might be required if data access is restricted. Tabletop displays have been found to encourage group members to work together in more cohesive ways, whereas wall displays are beneficial if information has to be discussed with a larger group of people [76].

Input: In the common desktop setup, input is provided for one person through one keyboard and one mouse. To support collaboration, ideally, each person would have at least one means of input. In addition, it would be helpful if this input was identifiable, making it possible to personalize system responses. If a collaborative system supports multi-user input, the access to a shared visualization and data set has to be coordinated. Also, synchronous interactions on a single representation may require the design and implementation of new types of multi-focus visualizations. Ryall et al. [77] have examined the problem of personalization of parameter changes for widget design, allowing widgets to be dynamically adapted for individuals within a group. Similar ideas could be implemented for personalization of information visualizations during collaborative work.

Resolution: Resolution is an issue both for the output (the display) and for the input. The display resolution has a great influence on the legibility of information visualizations. Large display technology currently often suffers from relatively low display resolution so that visualizations might have to be re-designed so that readability of text, color, and size not affected by display resolution. Also, large interactive displays are often operated using fingers or pens which have a rather low input resolution. Since information visualizations often display large data sets with many relatively small items, the question of how to select these small items using low input resolution techniques becomes an additional challenge that needs special attention [48].

3.3   Creating a Collaborative Environment

The key characteristics of co-located synchronous interactions as described by Olson and Olson [70] will apply to information visualization scenarios designed to support co-located collaboration. These characteristics include: a shared local context in which participants can interact with work objects, rapid feedback, and multiple channel information exchange (voice, gesture, etc.), and visibility of others' actions. These characteristics are further specified in the mechanics of collaboration [73], which describe basic operations of teamwork, or small scale actions and interactions that help people solve a task as a team. These mechanics apply to a variety of group and task settings. This section is discussed under the two major groupings of the mechanics of collaboration—communication and coordination—and under those issues relating to supporting varying collaboration styles.

Communication: Communication is an important part of successful collaborations. People need to be able to trigger conversations, communicate their intentions, indicate a need to share a visualization, and to be generally aware of their team members' actions. Group members need to be informed that some parameter of a shared display might have changed while they were busy working with an information visualization in a different part of the workspace. There has been considerable discourse on the importance of all team members being aware of all other team members' actions. Pinelle et al. [73] make a distinction between explicit and implicit communications. The design of support techniques for both types of communication needs to respect common social and work protocols [70]. For example, the interface should not require a group member to reach into or across another person's workspace in order to acquire or share visualizations or controls.

Explicit Communication: Enabling direct exchange of information through many channels such as voice, gestures, and deictic references facilitates collaborative work in general [73] as well as co-located collaboration [70]. It has been shown that the ability to annotate data and share insights in a written way is an essential part of the discovery process in distributed information visualization settings [46]. This collaborative need for annotation exists in traditional use of pen and paper based information as was observed in a study of teams working on information analysis tasks in a shared setting [68]. However, in digital systems messages of all types, written, voice, etc. might not always be as easily shared and how best to support this will require further research.

Implicit Communication: In co-located non-digital collaboration people are accustomed to gathering implicit information about team members' activities through such things as body language, alouds, and other consequential communications. This is an active research area in distributed collaboration research since the co-located evidence does not naturally become distributed. Co-located collaboration benefits from many of the co-present advantages, however, issues still arise. Some examples include: digital actions are not always readily visible (cursors are hard to see on large screens), menu actions can affect a remote part of the screen, as well as the general problems of change awareness [74]. Thus while implicit communications do support awareness in a co-located setting already to some extent, some system changes made by a collaborator can still remain unnoticed if the collaborative system does not provide appropriate feedthrough (i. e. a reflection of one person's actions on another person's view). In collaborative information visualization, for example, it might be important to consider appropriate awareness for operations that make changes to the underlying dataset.

Imagine a co-located system in which each collaborator works in parallel on a different view using a different file-system representation. If one collaborator discovers an old version of a file and decides to delete it (a value operation [23]), this change might go unnoticed if the other person is looking at a view of the data that does not include the current file or it might be completely surprising to the other person to see a file in their representation disappear. Some research has proposed policies to restrict certain members from making unsuspected global changes to a dataset [75]; however, while earlier research on information visualization discussed the differences between view and value operators (e. g. [23]), most recent research in multiple-view visualization tends to favour view operations (filtering of unwanted data rather than deletion). This seems likely to be most appropriate during collaboration.

It has also been shown the location and orientation of artifacts is used to support implicit communication in non-digital settings providing information on such things as who is working with which artifacts and when one person wants to initiate communication about a particular artifact [53] and that this translates to digital settings [54]. This consideration, providing for artifact mobility and freedom orientation, will probably also be important in supporting information visualization collaboration.

Coordination: In group settings, collaborators have to coordinate their actions with each other. Here, we describe several guidelines for how to support the coordination of activities in collaborative information visualization applications.

Workspace Organization: Typical single-user information visualization systems impose a fixed layout of windows and controls in the workspace. Previous research has shown that, on shared workspaces, collaborators tend to divide their work areas into personal, group, and storage territories [81]. This finding implies that a group interaction and viewing space is needed for collaborative data analysis where the group works on a shared representation of the data or in which they can share tools and representations. Also, the possibility of exploring the data separately from others, in a personal space, is necessary. Flexible workspace organization can offer the benefit of easy sharing, gathering, and passing of representations to other collaborators. By sharing data in the workspace, representations will be viewed by team members with possibly different skill sets and experiences and, therefore, subjected to different interpretations. Also, by being able to move and rotate representations in the workspace, an individual can gain a new view of the data and maybe discover previously overlooked aspects of the data display.

Collaborative information visualization systems should allow for social interaction around data displays [46]. If visualizations can be easily shared, team members with different skill sets can share their opinions about data views, suggest different interpretations, or show different venues for discovery. By offering mechanisms to easily rotate and move objects, comprehension, communication, and coordination can be further supported [54]. Rotation can support comprehension of a visualization by providing alternative perspectives that can ease reading and task completion, coordination by establishing ownership and categorizations, and communication by signaling a request for a closer collaboration [53]. By allowing free repositioning, re-orientation we can also make use of humans' spatial cognition and spatial memory and possibly better support information selection, extraction, and retrieval tasks [68]. Mechanisms for transfer and access to information visualization in the workspace should be designed in a way that they respect common social work protocols [53,81].

Changing Collaboration Styles: Tang et al. [90] describe how collaborators tend to frequently switch between different types of loosely and closely coupled work styles when working over a single, large, spatially-fixed information display (e. g., maps or network graphs). A study by Park et al. [72] in distributed CAVE environments discovered that, if the visualization system supports an individual work style, users preferred to work individually on at least parts of the problem. For information visualization systems, an individual work style can be supported by providing access to several copies of one representation. The availability of unlimited copies of one type of representation of data allows group members to work in parallel. More closely coupled or joint work on a single view of the data can be supported by implementing the possibility of concurrent access and interaction with the parameters of an information visualization. Free arrangements of representations also support changing work styles. Representations can be fluidly dragged into personal work areas for individual or parallel work and into a group space for closer collaboration.

3.4   Designing Information Visualizations for Co-located Collaboration

Many known information visualization guidelines still apply to the design of information visualizations for co-located collaborative use (e. g., [10,92,99]). In this section, we discuss changes and additions to aspects that need to be considered when designing information visualizations for co-located collaborative settings. Thus, much of this discussion simply delineates research questions that may of specific interest when designing information visualizations to support co-located collaboration.

Representation Issues: Spence [87] defines representation as "the manner in which data is encoded," simplifying Marr's [62] definition of representation as a formal system or mapping by which data can be specified. The concept of representation is core to information visualization since changes in representations cause changes in which types of tasks are most readily supported. As in Marr's [62] example, the concept of thirty-four can be represented in many ways. To look at three of them; Arabic numerals, 34, ease tasks related to powers of ten; Roman numerals, XXXIV, simplify addition and subtraction; and a binary representation, 100010, simplifies tasks related to powers of two. Not surprisingly, Zhang and Norman [110] found that providing different representations of the same information to individuals provides different task efficiencies, task complexities, and changes decision-making strategies. Questions arise as to what are the most effective representations during collaboration. Will certain representations be better suited to support small group discussions and decision making? Will multiple representations be more important to support different people's interpretation processes? Will new encodings or representations be needed for collaborative work scenarios? Appropriate representations might have to be chosen and adapted depending on the display type chosen but whether completely new designs are required is not yet clear.

For example, different representations may have to be accessible in an interface because in a collaborative situation, group members might have different preferences or conventions that favour different types of representations. Gutwin and Greenberg [39] have discussed how different representations of the workspace affect group work in a distributed setting. They point out that providing multiple representations can aid the individual but can restrict how the group can communicate about the objects in the workspace. This extends to co-located settings, in which several representations of a dataset can be personalized according to taste or convention, making it harder to relate individual data items in one representation to a specific data item in another. For example, relating one specific node in a treemap [50] to another node in a node-link diagram might require a search to locate the respective node in the other representation. Implementing mechanisms to highlight individual data items across representations might aid individuals when switching between group and parallel data exploration.

Findings suggest that the availability of multiple, interactively accessible representations might be important for information visualization applications since the availability of multiple data representation can change decision making strategies [52]. Also differing representations have an influence on validation processes in information analysis [79], and more easily support people working in parallel on information tasks [72]. While this is probably applicable, empirical evidence directly linking these finding to collaborative information visualization has not yet been gathered.

It is also possible that the actual mappings used in representations may have to be re-thought. For example, spatiality or the use of position/location is commonly an important aspect of representation semantics. However, spatiality as manifested in territoriality is a significant factor for communication and coordination of small group collaboration. It is an open question as to whether there is a trade-off between these two uses of spatiality.

Presentation Issues: Presentation has been defined as 'something set forth for the attention of the mind' [63] and as 'the way in which suitably encoded data is laid out within available display space and time' [87]. From these definition is clear that changing display configurations, as is usually the case to support co-located collaboration, will impact the types of presentations techniques that are possible and/or appropriate. Common presentation techniques include pan & zoom, focus & context, overview & detail, filtering, scrolling, clutter reduction, etc.

A common theme in information visualization is the development of presentation techniques that overcome the problem of limited display space (e. g. [4,20,49]). In collaborative scenarios, information visualizations might have to cover larger areas than in a single user scenario as group members might prefer to work in a socially acceptable distance from each other. The display space might also have to be big enough to display several copies of one representation if team members want to work in parallel.

If groups are working over a shared presentation of data, presentations might have to be adapted to allow collaborators to drill down and explore different parts of the data in parallel. Collaborative information visualizations will likely have to sup-port multiple simultaneous state changes. This poses additional problems of information context. Team members might want to explore different parts of a dataset and place different foci if the dataset is large and parts of the display have to be filtered out. Information presentations might have to be changed to allow for multi-focus exploration that does not interfere with the needs of more than one collaborator. For example, DOI Trees [18] or hyperbolic trees [55] are examples of tree visualizations in which only one focus on the visualization is currently possible. ArcTrees [67] and TreeJuxtaposer [65], for example, allow for multi foci over one tree display but these were not designed to take the information needs of multiple collaborators into account and might still occlude valuable information.

An example for visualization presentation changes based on a collaborative circular tabletop environment has been presented in [94]. The presentation of the circular node-link tree layout was modified to rotate all nodes towards the boundary and a "magnet" was implemented to rotate nodes towards just one team member. Nodes were also changed in size; as leaf nodes were placed closer towards team members, in their personal space [81], they were decreased in size and the nodes towards the center of the table were enlarged to allow for easier shared analysis of the node contents in the group space [81]. A possible extension of this work is to think about placing and re-arranging nodes automatically based on the placement and discovery interests of team members or based on the individual or shared discoveries that have been made.

The presentation of visualizations might also have to take available input devices on a shared large display into account. If fingers or pens are used as an input device, the selection might not be accurate enough to select small information items. A common task in information visualization is to re-arrange data items (e. g. by placing points of interest), to request meta-information [85] (e. g. by selecting an item), or to change display parameters by selecting an item. If the displayed dataset is large, it often covers the full screen and reduces individual items to a few pixels. Previous research has attempted to solve the issue of precise input for multi-touch screens (e. g. [8]) but they might not be applicable if the whole visual display is covered with items that can possibly be selected. Alternatively, information presentations could be changed to allow for easier re-arrangement and selection of items, for example, with lenses [20]. DTLens [35] presents an initial exploration of the use of lenses in co-located collaboration.

The resolution of a large display has an influence on the legibility of data items. It is known that the reading of certain visual variables is dependent on the size and resolution in which they are displayed [99]. Information visualizations also often rely on textual labels to identify data items which may be hard to read on low-resolution displays. The presentation size of individual items and labels may have to be adapted to compensate for display resolution.

View Issues: The term view is common in information visualization literature and view operations (changing what one currently sees) have been defined as distinct from value operations (changing the underlying data) [23], however, this use of the term view also incorporated changes in visual aspects of representation, presentation. Blurring the distinction between view and presentation changes has not been problematic because with a single viewer and a single display these are often concurrent. A change in view can be simply looking at exactly the same presentation and representation of the same data merely from a different angle or it can include changes in all three factors.

In a co-located collaborative setting, of necessity there are as many views of a given presentation as there are people in the group. Also since collaboration practices often include mobility, a given person's view will change as they move in the physical setup. This factor has recently begun to receive attention in the CSCW community. Nacenta et al. [66] have shown that righting (orienting a piece of 2D information into the proper perspective, by means of motion tracking, really aids comprehension. Hancock and Carpendale [40] consider the same problem for horizontal displays looking for non-intrusive interactive solutions. Since a study by Wigdor et al.[105] has indicated that angle of viewing affects readability of certain visual variables; this issue will be an important one for collaborative information visualizations. This research on how view-angle distortion affects perception in a single and multi-display environment suggests that certain types of representations may need to be modified in order to be used on a digital tabletop display and that information visualizations should not be compared across multiple display orientations. However, as visual variables were tested in isolation (e. g. length, direction, only) further evaluations have to be conducted to see whether participants will correct for possible distortion if the variables are presented in conjunction with others or whether view correction [66,40] might compensate.

Visualizations that can be read from multiple angles and orientations (e. g. circular tree layouts vs. top-down layouts) might be more appropriate for display on a horizontal surface. However, it is not clear whether participants would try to read oriented visualizations upside down and make wrong conclusions based on these readings or whether they would simply re-orient the visualization to correct the lay-out. Observations of collaborative information visualization scenarios point in the latter direction [68].

Gutwin and Greenberg [39] discuss issues about viewing a representation in relation to distributed scenarios. However, parallels can be drawn to co-located scenarios in which collaborators work with multiple linked copies of the same representation of a dataset. These essentially represent multiple movable "split viewports" [39]. The suggested solutions for distributed settings include radar views, overview+detail solutions, and cursor eye-view. Whether the benefit of these solutions in a co-located setting outweighs the possible distractions they might create, however, would have to be evaluated. Further guidelines for using multiple views in information visualization can be found in [5] and provide a starting point for tailoring multiple views for collaborative visualization.

Issues of view may develop in collaborative information visualization settings if collaborators want to switch from loosely-coupled to closely-coupled workstyles [90] and share discoveries they have made with the other group members. If one collaborator worked with different view of the dataset it might be difficult to locate the information in the other persons' view. Another important factor to consider when developing a collaborative viewing strategy is the establishment of territories [81] for personal, group and storage purposes that is suggested as beneficial for group coordination (see "Coordination" above),

A study by Yost and North [109] compared the ability of visualizations to display large amounts of data normalized across either a small or a large high-resolution vertical display. Their study showed that the visualizations used were perceptually scalable but that people preferred different visualizations on the large vs. small display, as some were found to be easier to read than others depending on the screen size. How these preferences would change for collaborative work would have to be evaluated.

Interaction Issues: Most interaction issues deal with interaction with representations, presentations and views, thus discussing them here would overlap with points raised under these headings. However, there are some more general interaction issues. When people are co-located, they are in the situation in which people naturally collaborate, the situation in which people have collaborated for centuries. When face-to-face, people naturally know how to collaborate and are so used to picking up subtle cues from each other that they may do this without even being conscious of the precise details of the underlying coordination and communication practices that are in play. As the developers of co-located collaborative information visualizations, our task is to facilitate information access and exploration without interfering with the social protocols that make collaboration effective. However, to do this we have to understand what these social collaboration practices are and specifically if there are any differences when people are collaborating using visual information. Some factors are:

Interactive Response Rates: Information visualization has always had a lot of requirements in that it deals with extremely large and complex data sets and in that it can have considerable graphics requirements for these complex representations. Adding larger screens, more screens, higher pixel counts, multiple simultaneous inputs, and possibly multiple representations will increase computational load adding more requirements to the challenge of maintaining good interactive rates. Thus implementations of collaborative information visualizations will have to be carefully designed for efficiency. While continued hardware advances will mitigate this to some extent, it will be important to address issues in both efficient data processing and fast graphic rendering.

Interaction History: A history task has been defined as a task that involves keeping a history of actions to support undo, replay, and progressive refinement [85]. In a collaborative scenario keeping such a history can have other benefits. If a visualization tracks and reveals which data items have been visited and by whom this information could be valuable for collaborators helping them understand their team members' actions, find unexplored parts of a visualization or to confirm discoveries made by others. A visualized interaction history may support collaboration by promoting mutual understanding of team members involvement in the task [24] and may help keep group members aware of each others actions as people shift from individual to shared views of the data [39]. An exploration history can be useful in such activities as validating work done, in explaining a discovery process to other team members, and in supporting discussions about data explorations.

Information Access: Exactly how to handle information access is an important collaboration issue. The main themes in the research discussion thus far have been motivated by social protocol issues and data centric concerns. While these have not been seen as mutually exclusive they are quite distinct ideas. The social protocol theme has made considerable use of observational studies to better understand exactly what are the social protocols and how do they impact collaboration. These understandings are then used as a basis for software design.

The data centric approach discusses factors such as who has (or does not have) rights to which parts of the data?, who can change the scale, zoom, or rotation settings for a shared view of the data? And how does a data item get passed between team members (hand-off). Restriction has been suggested as a means to stop certain members from making unsuspected global changes to the data that might change other members' view of the same data [75]. Similar issues pertaining to workspace awareness (individual vs. shared views), artefact manipulation (who can make which changes), and view representation have been raised [39]. Is a single shared representation adequate? Should a system allow for multiple representations? Should the exploration on multiple representations of the same dataset be linked or be completely independent?

Fluid Interaction: The fluidity of interactions in a shared workspace influences how much collaborators can focus on their task rather than on the manipulation of interface items [82]. This implies that in a collaborative information analysis scenario, parameter changes to the presentation or representation of a dataset should require manipulation of as few interface widgets (menus, slider, etc.) as possible and little or no changes of input modalities (mouse, keyboard, pen, etc.). A study on collaborative information visualization systems has similarly reported that groups worked more effectively with a system in which the required interactions were easier to understand [61]. This poses a challenge to information visualization tool designers as typically a high number of parameters are required in visualization systems to adapt to the variability in dataset complexity, size, and user tasks.

4   Collaborative Visualization on the Web

Visual analysis is rarely a solitary activity. A business analyst may notice an unexpected trend in a chart of sales figures – but then she's likely to confer with a colleague, who may share the chart with a manager, who later might present it to executives. Such scenarios of collaboration and presentation across both time and space are common in business and scientific visualization. Just as a good visualization takes advantage of the power of the human visual system, it can also exploit our natural social abilities. Accordingly, designers of visualization systems should consider not only the space of visual encodings, but mechanisms for sharing and collaboration. At minimum, systems should enable people to communicate about what they see so they can point out discoveries, share knowledge, and discuss hypotheses.

The social aspects of visualization have taken on new importance with the rise of the Web. While collaboration in small groups remains ubiquitous, it is now also possible for thousands of people to analyze and discuss visualizations together. These scenarios are driven by the fact that users can interact remotely from anywhere on the globe and access the system at different times. Partitioning work across time and space holds the potential for greater scalability of group-oriented analysis. For example, one decision making study found that asynchronous collaboration resulted in higher-quality outcomes – broader discussions, more complete reports, and longer solutions – than face-to-face collaboration [6].

Web-based collaboration around visualizations introduces new challenges for research, as most work on collaborative visualization has been done in the context of synchronous scenarios: users interacting at the same time to analyze scientific results or discuss the state of a battlefield. As described in the previous section, co-located collaboration usually involves shared displays, including large wall-sized screens and table-top devices (e. g., [28,31]). Systems supporting remote collaboration have primarily focused on synchronous interaction [3,14]), such as shared virtual workspaces (e. g., [1,24]) and augmented reality systems that enable multiple users to interact concurrently with visualized data (e. g., [25,9]). In addition, the increasing availability of table-top and large public displays has prompted researchers to experiment with asynchronous, co-located visualization (same place, different time), often in the form of ambient information displays (e. g., [21]).

In this section, we instead focus on the kind of collaboration that is most common over the Web: remote collaboration across time and space. Our goal is to summarize the work done to date and indicate promising research directions. We first review recent web-based systems supporting social data analysis around visualizations, highlighting the collaborative features provided by these systems and how they have been used in practice. We then discuss a number of outstanding challenges for asynchronous collaborative visualization and identify avenues for future research.

4.1   Web-Based Collaborative Visualization Systems

Though web-based collaboration around visualizations is still in its infancy, a handful of commercial and research systems in this area have recently been introduced. Here we discuss contemporary visualization systems that support asynchronous collaborative analysis (shown in Fig. 2), documenting the collaborative features supported by these tools and initial reports of their usage.

DecisionSite Posters: Adding Collaboration to a Single-User Tool: DecisionSite Posters is a feature of the Spotfire product sold by TIBCO, Inc. Users of Spotfire's desktop-based visualization system can capture snapshots of their analyses and publish them on an intranet as "posters." View sharing is supported, as each poster has a unique URL than can be easily distributed. Each poster also sup-ports unthreaded text comments on a side panel. However, posters do not allow annotations, limiting the ability of collaborators to point at specific trends or outliers.

As described in [96], the communication capabilities in DecisionSite Posters have been used in an unexpected way. Instead of engaging in complex conversations by using the comment panel, as envisioned by the system's designers, users have largely used the tool for presenting their findings to colleagues. The ability

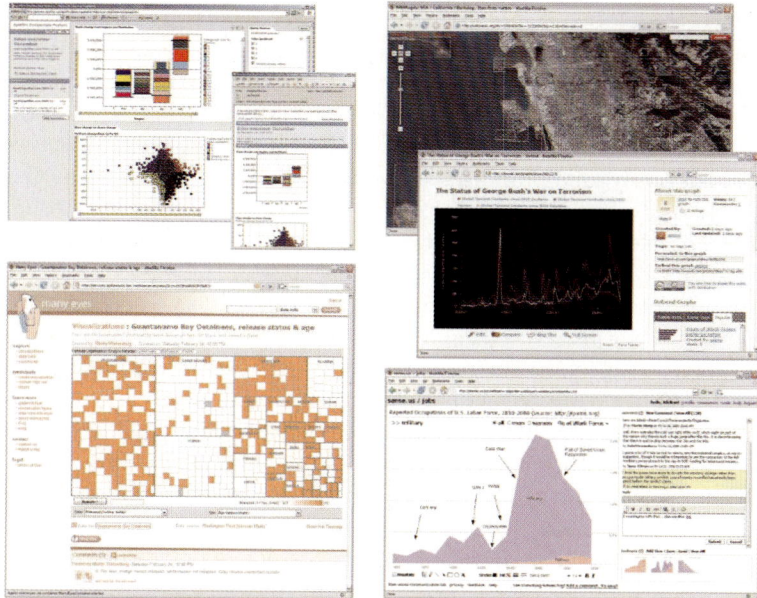

Fig. 2. Asynchronous Collaborative Visualization Systems. Clockwise from top-left: Spotfire DecisionSite Posters, Wikimapia, Swivel, sense.us, and Many Eyes.

to create comments with pointers into the visualization provides an easy way to choreograph a step-by-step presentation.

Swivel: Sharing Data on the Web: Swivel.com is a web site that supports sharing and discussion around data. The service appears to be modeled on sites such as YouTube that support sharing of other media. In keeping with this model, Swivel allows users to upload data sets and talk about them in attached discussion forums. In addition, the site automatically generates graphs by combining columns from uploaded data sets into bar charts, pie charts, and scatter plots. Pointing behavior on the site appears limited.

Although the graphs on Swivel are not interactive, the site provides an example of social data analysis in action, in particular the importance of collaborative publishing and sharing of visualizations. While there do not seem to be many extensive conversations in Swivel's discussion area there has been significant use of Swivel's graphs among bloggers to discuss statistics. In other words, it appears that the ability to publish graphs for use in other contexts is most valuable to Swivel's users.

Wikimapia: Collaborative Geographic Annotation: Wikimapia.org is a web site enabling collective annotation of geographic satellite imagery, and is representative of similar efforts such as Google Earth and mash-ups created with web APIs to mapping services. The site provides a zoomable browser of satellite photos across the globe, along with the ability to select geographic regions for annotation with names and additional data (Fig. 2). View sharing is supported through automatically updating URLs. As the view is panned or zoomed, the current URL updates dynamically to reflect the current zoom level and latitude and longitude values. Pointing is supported through annotations. Users can draw rectangular and polygonal annotations, which scale appropriately as the map is zoomed. To avoid clutter, annotations are filtered as the view is zoomed; the viewer does not see annotations that are too small to be legible or so large they engulf the entire display, improving the scalability of the system.

Wikimapia supports conversation using an embedded discussion technique. Each annotation is a link to editable text. Descriptive text about a geographic region can then be edited by anyone, similar to articles on Wikipedia. Discussion also occurs through voting. When annotations are new, users can vote on whether they agree or disagree with the annotation. Annotations that are voted down are removed from the system. For instance, the small town of Yelapa, Mexico is located on an inlet in a bay near Puerto Vallarta. However, the bay has a number of inlets very close together. As a result, multiple conflicting annotations for Yelapa appeared. Through voting, the incorrect regions were discarded and the correct annotation was preserved.

sense.us: Social Data Analysis of U.S. Census Data: Sense.us is a proto-type web application for social visual data analysis [46]. The site provides inter-active visualizations of 150 years of United States census data, including stacked timelines, choropleth maps, and population pyramids. With a URL-based book-marking mechanism, it supports collaboration through doubly-linked discussion, graphical annotations, bookmark trails, and searchable comment listings.

Discussion occurs via a doubly-linked conversation model. Searchable com-ment listings provide links back into the visualization, while navigating in a visu-alization automatically causes related comments to be retrieved for the current view. By tying commentary directly to specific view states, comments become less ambiguous, enabling remarks such as "that spike" or "on the left" to be more easily understood. Pointing occurs through freeform graphical annotations and view sharing is facilitated by URLs that automatically update as users navigate the visualizations. Sense.us also allows users to collect view links in a "bookmark trail" of view thumbnails. Users can then drag-and-drop view thumbnails from the trail into a comment text field, thereby adding a hyperlink to the saved view within the comment. In this way, users can provide pointers to related views and create tours through the data.

Studies of the sense.us system revealed interesting patterns of social data analysis. Users would make observations about the data, often coupled with questions and hypotheses about the data. These comments often attracted fur-ther discussion. For example, within a visualization of the U.S. labor force over time, a spike and then decline in the number of dentists prompted discussion ranging from the fluoridation of water to stratification within the dentistry pro-fession, with a rise in the number of hygienists corresponding to the decline of dentists. There was also an interesting interaction between data analysis and social activity. Users who tired of exploring visualizations turned their focus to the comment listings. Reading others' comments sparked new questions that led users back into the visualization, stimulating further analysis. The sense.us prototype was initially available on a corporate intranet which provided employees with blogs and a social bookmarking service. Users of sense.us found ways to publish their findings, typically by taking screenshots and then placing them on blogs or the bookmarking service with application bookmarks. These published visualizations drew additional traffic to the site.

Many Eyes: Web-Based Visualization and Publishing: Many-Eyes.com [97] is web-based service that combines public data sharing with interactive visualizations. Like social data analysis sites such as Swivel, site members can upload data sets and comment on them. Unlike Swivel, however, Many Eyes offers a palette of interactive visualization techniques – ranging from bar charts to treemaps to tag clouds – that visitors may apply to any data set. Users may post comments on the visualizations, including bookmarks for particular states.

The pointing and discussion capabilities of Many Eyes are used in a variety of ways. The site contains some lengthy conversations around visualizations, although the great majority of visualizations have no comments. One class of visualizations, however, did lead to lengthy onsite discussions: visualizations that sidestepped sober analysis and were instead playful or comical. One person, for example, initiated a game based on a visualization of Shakespearean poetry in which he used the highlighting mechanisms to pick out alphabetically ordered words to make pseudo-Elizabethan epithets. These games frequently attracted many "players."

The Many Eyes site also can be viewed as a publishing platform, since the visualizations that users create are publicly visible and may be linked to from other web pages. Many bloggers have taken advantage of this, and perhaps as a result the deepest analyses of Many Eyes visualizations have occurred as part of blog entries that reference the site. In one example, a blog at the Sunlight Foundation (a political reform organization) published a Many Eyes tag cloud to analyze messages between the U.S. Congress and the U.S. Department of Defense. The blog entry framed the results as a funny-but-sad surprise: the most common phrases had nothing to do with current pressing issues, but rather requests for congressional travel. In another case, a user created a visualization of the "social network" of the New Testament. Not only was this visualization linked to from more than 100 blog entries, but another user went to the trouble of recording a YouTube video of himself interacting with the visualization and narrating what he learned. These phenomena again underscore the importance of publishing mechanisms for collaborative visualization.

Summary: The previous examples of web-based collaborative visualization present a number of common design decisions, but also some important differences. All systems support view sharing through URL bookmarks and enable discussion through text comments. Furthermore, usage examples from these systems suggest that users derive great value from being able to share and embed visualizations in external media such as blogs. One salient difference between systems is the varied forms of pointing within visualizations: selecting individual items in Many-Eyes, creating polygonal geographic regions in Wikimapia, and drawing freeform graphics in sense.us. Another difference is the way commentary is attached to visualization views. Spotfire Decision Site Posters, Swivel, and Many Eyes all support blog-style unthreaded comments for individual visualizations. In contrast, Wikimapia supports commentary attached to geographic annotations, while sense.us provides threaded comments tied to specific states of the visualization and retrieved dynamically during exploration.

4.2   Research Issues in Web-Based Collaborative Visualization

As described in the previous section, developers of collaborative visualization systems face design decisions of how to support discussion, annotation, and integration with external services. Future research in asynchronous collaborative visualization needs to provide guidance through this design space, as well as develop novel techniques for better facilitating collaborative analysis. In this section, we identify five areas in which we expect additional research to make important contributions to improving the state-of-the-art: the structure and integration of collaborative contributions; engagement and incentives; coordination and awareness; pointing and reference; and presentation, dissemination, and story-telling.

Structuring and Integrating Contributions: A fundamental aspect of successful collaboration is an effective division of labor among participants. This involves both the segmentation of effort into proper units of work and the allocation of individuals to tasks in a manner that best matches their skills and disposition. Primary concerns are how to split work among multiple participants and meaningfully aggregate the results.

Drawing on examples such as online discussions, open source software, and Wikipedia, Benkler [7] introduces the concepts of modularity, granularity, and cost of integration in the peer production of information goods. Modularity refers to how work is segmented into individual units of contribution, while granularity refers to the scope of these units and how much effort they require. For example, in online scenarios where incentives tend to be small and non-monetary, a small granularity may encourage people to participate in part due to the ease of contributing. The cost of integration refers to the effort required to usefully synthesize contributions into a greater whole. Collaborative work will only be effective if the cost of integration is low enough to warrant the overhead of modularization while enforcing adequate quality control. There are a number of mutually inclusive approaches to handling integration: automation (automatically integrating work through technological means), peer production (casting integration as an additional collaborative task given to trusted participants), social norms (using social pressures to reduce vandalistic behavior), and hierarchical control (exercising explicit moderation).

Collaborative visualization can similarly be viewed as a process of peer production of information goods. Stages in this process include uploading data sets, creating visualizations, and conducting analysis. To support this process, it is important to identify the specific forms of contribution (modules) that users might make and how to integrate these contributions. Existing frameworks for aiding this task include structural models of visualization design and sensemaking processes [17]. As shown in Fig. 3, each of these models suggests tasks that contribute to collaborative analysis, including data cleaning, moderation, visualization specification, sharing observations, positing hypotheses, and marshaling evidence. These concerns are given further treatment in [43].

Once modules have been identified, one can then attempt designs which reduce the cost structure of these tasks. Consider the issue of scale. Most of the examples in the previous section use sequential text comments to conduct analytic discussion. However, it is unclear how well this form of communication will scale to massive audiences. An open research problem is the creation of new forms of managed conversation that have a lower cost of integration, enabling people to understand and contribute to analysis without having to wade through hundreds of individual comments. For example, Wikipedia relies on human editing coupled with a revision management system to integrate and moderate contributions. Alternatively, systems with highly structured input such as NASA ClickWorkers [7] or von Ahn's (2006) "games with a purpose" [98] rely on purely automated techniques. Some middle ground between these approaches should be possible for collaborative analysis, such as argumentation systems that model hypotheses and evidence as first class objects. One example of such a system is CACHE [11], which maintains a matrix of hypotheses and evidence, with collaborators providing numerical measures of the reliability of evidence and the degree to which evidence confirms or disconfirms the hypotheses. These scores can then be averaged to form a group assessment. Other possibilities include augmenting graphical workspaces such as the Analysis Sandbox [107] with collaborative authoring features or automatic merging of representations (c.f., [13]).

Engagement and Incentives: If collaborators are professionals working within a particular context (e. g., financial analysts or research scientists) there may be existing incentives, both financial and professional, for conducting collaborative work. In a public goods scenario, incentives such as social visibility or sense of contribution may be motivating factors. Incorporating incentives into the design process of collaborative visualization systems may increase the level of user contributions, and could even provide additional motivation in situations that already have well established incentive systems.

Benkler posits an incentive structure for collaborative work consisting of monetary, hedonic, and social-psychological incentives [7]. Monetary incentives refer to material compensation such as a salary or cash reward. Hedonic incentives refer to well-being or engagement experienced intrinsically in the work. Social-psychological incentives involve perceived benefits such as increased status or social capital.

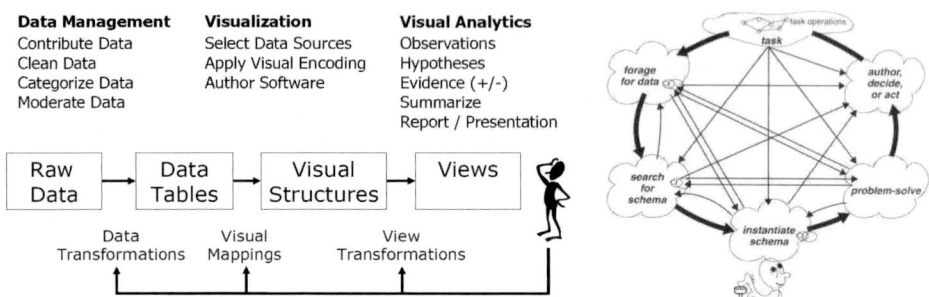

(a) Collaborative activity might be introduced at any phase of the information visualization pipeline.

(b) The sensemaking model in [17] can be applied to identify potential mechanisms for collaborative analysis (e. g., [43])

Fig. 3. Models of the Visualization Process.

Observations of social use of visualization have noted that visualization users are attracted to data which they find personally relevant [45,96,100]. For example, in collaborative visual analysis of the occupations of American workers [46]), users often search for their own profession and those of their friends and family, similar to how people search for names in the popular NameVoyager visualization [100]. The hypothesis is that by selecting data sets or designing their presentation such that the data is seen as personally relevant, usage rates will rise due to increased hedonic incentive. For example, geographic visualizations may facilitate navigation to personally relevant locations through typing in zip codes or city names, while a visualization of the United States' budget might communicate how a specific user's taxes were allocated rather than only listing total dollar amounts.

In the case of social-psychological incentives, the visibility of contributions can be manipulated for social effects. Ling et al [56] found that users contributed more if reminded of the uniqueness of their contribution or if given specific challenges, but not under other theoretically-motivated conditions. Cheshire [22] describes a controlled experiment finding that, even in small doses, positive social feedback on a contribution greatly increases contributions. He also found that visibility of high levels of cooperative behavior across the community increases contributions in the short term, but has only moderate impact in the long term. These studies suggest that social-psychological incentives can improve contribution rates, but that the forms of social visibility applied have varying returns. One such incentive for visual analysis is to prominently display new discoveries or successful responses to open questions. Mechanisms for positive feedback, such as voting for interesting comments, might also foster more contributions.

Finally, it is worth considering game play as an additional framework for increasing incentives. In contrast to environments such as spreadsheets, many visualizations already enjoy game-like properties, being highly visual, highly interactive, and often animated. Heer [42] discusses various examples in which playful activity contributes to analysis, applying insights from an existing theory of playful behavior [16] that analyzes the competitive, visceral, and teamwork building aspects of play. For example, scoring mechanisms could be applied to create competitive social-psychological incentives. Game design might also be used to allocate attention, for example, by creating a team-oriented "scavenger hunt" analysis game focused on a particular subject matter. Salen and Zimmerman [78] provide a thorough resource for the further study of game design concepts.

Coordination and Awareness: An important aspect of collaborative action is awareness of others' activities, allowing collaborators to gauge what work has been done and where to allocate effort next [29,19]. Within asynchronous contexts, participants require awareness of the timing and content of past actions. This suggests that designs should include both history and notification mechanisms (e. g. [15]) for following actions performed on a given artifact or by specific individuals or groups. Browseable histories of past action are one viable mechanism, as are subscription and notification technologies such as RSS (Really Simple Syndication).

User activity can also be aggregated and abstracted to provide additional forms of awareness. Social navigation [30] involves the use of activity traces to provide additional navigation options, allowing users to purposefully navigate to past states of high interest or explore less-visited regions (the "anti-social navigation" of Wattenberg & Kriss [100]). For example, navigation cues may be added to links to views with low visitation rates or to action items such as unanswered questions and unassessed hypotheses. One recent study [106] provides evidence that social navigation cues can simultaneously promote revisitation of popular or controversial views while also leading to a higher rate of unique discoveries. Future research is needed to further develop and evaluate other forms of awareness cues for supporting collaborative analysis.

Pointing and Reference: When collaborating around visual media, it is common for one to refer to visible objects, groups, or regions [26,12]. Such references may be general ("north by northwest"), definite (named entities), detailed (described by attributes, such as the "blue ball"), or deictic (pointing to an object and saying "that one"). Hill and Hollan [47] discuss the various roles that deictic pointing gestures can play, often communicating intents more complicated than simply "look here". For example, different hand gestures can communicate angle (oriented flat hand), intervals (thumb and index finger in "C" shape), groupings (lasso'ing a region), and forces (accelerating fist). While other forms of reference are often most easily achieved through speech or written text, deictic reference in particular offers important interface design challenges for collaborative visualization. Nuanced pointing behaviors can improve collaboration by making it easier to establish the object of conversation. Hill and Hollan argue for "generally applicable techniques that realize complex pointing intentions" by engaging "pre-attentive vision in the service of cognitive tasks."

A standard way to point in a visualization is brushing: selecting and high-lighting a subset of the data. Naturally, these selections should be sharable as part of the state of the visualization. In addition, a palette of visual effects richer than simple highlighting can let users communicate different intents. For exam-ple, following time-varying values of selected items in a scatter plot is easier when the selected items leave trails as they move over time. The selected items and their trails are even more salient if non-selected items are simultaneously de-emphasized. Brushing-based forms of pointing have the advantage that the pointing action is tied directly to the data, allowing the same pointing gesture to be reapplied in different views of the same data. As "data-aware" annotations are machine-readable, they can also be used to export subsets of data and help steer automated data mining [108].

Freeform graphical annotations are a more expressive method of pointing in visualizations. Drawing a circle around a cluster of items or pointing an arrow at a peak in a graph can direct the attention of remote viewers; at the same time, the angle of the arrow or shape of the hand-drawn circle may communicate emo-tional cues or add emphasis. However, while such drawing and vector graphic annotations allow a high degree of expression, they only apply to a single view in the visualization, without any explicit tie to the underlying data. Freeform annotations can persist over purely visual transformations such as panning and zooming, but they are not data-aware and may become meaningless in the face of data-oriented operations such as filtering or drill-down. A promising research direction is hybrid approaches that combine aspects of both brushing and graph-ical annotation. The resulting techniques could create graphical annotations that are tied to data points so that they can be reapplied in other views of the data.

Presentation, Dissemination, and Story-Telling: Common forms of infor-mation exchange in group sensemaking are reports and presentations. Narrative presentation of an analysis "story" is a natural and often effective way to commu-nicate findings, and has been observed as a primary use of Decision Site Posters. Furthermore, usage of Swivel, sense.us, and Many Eyes leverages external me-dia such as blogs and social bookmarking services as additional communication channels in which to share and discuss findings from visualizations. The challenge to collaborative visualization is to provide mechanisms to aid the creation and distribution of presentations. For example, sense.us [46] allows users to construct and share trails of related views to create tours spanning multiple visualizations and the GeoTime Stories [32] system supports textual story-telling with hyper-links to visualization states and annotations. However, neither system yet allows these stories to be exported outside the respective applications. In future work, such mechanisms could be improved with support to build presentations semi-automatically using interaction histories, export such presentations into external media, and apply previously discussed pointing techniques. A related issue is to enable follow-up analysis and verification for parts of the analysis story, enabling presentations to serve as a catalyst for additional investigation.

4.3 Summary

In this section, we introduce an emerging use of interactive visualization: collaborative visual analysis across space and time. The Web has opened up new possibilities for large-scale collaboration around visualizations and holds the potential for improved analysis and dissemination of complex data sets. A new class of systems explores these possibilities, enabling web-based data access, exploration, view sharing, and discussion around both static and interactive visualizations. Already, these systems exhibit the promise of web-based collaboration, providing examples of collective data analysis in which group members combine their knowledge to make sense of observed data trends and disseminate their findings.

Still, many research questions remain on how to structure collaboration. For example, how can we move beyond simple textual comments to better scale and integrate diverse contributions? Interested readers may wish to consult [96,46,43] for further discussions on this topic. As described in section 2, another open question is how to design for particular audiences. Different scenarios – including scientific collaboration, business intelligence, and public data consumption – involve different skill sets, scales of collaboration, and standards of quality. Going forward, case studies in these scenarios are crucial to better tailoring visualization tools to such varied audiences. By enabling users to collectively explore data, share views and findings, and debate competing hypotheses, the resulting collaborative visual analysis systems hold the potential to improve the number and quality of insights gained from our ever-increasing collections of data.

5 Conclusion

The adoption of visualization technologies by people from different walks of life has important implications for visualization research and development. Visualization construction tools are lowering barriers to entry, resulting in end-user created visualizations of every kind of data set imaginable. Concurrently, new technologies enabling collaborative use of visualizations in both physical and online settings hold the potential to change the way we explore, analyze, and communicate. In this paper, we have sought to identify these emerging trends and provide preliminary design considerations for advancing the state-of-the-art of visualization and visual analytic tools.

As a parting comment, we note that the release of visualization tools "into the wild" will undoubtedly result in a plethora of unexpected developments. Equipped with new creation and collaboration tools, users will almost certainly re-appropriate these technologies for unexpected purposes. Already, use of systems like Many-Eyes has revealed new genres of data-oriented play and self-expression that complement more traditional analytic activities.

As researchers, it is imperative that we interface with these developments in a productive fashion. It is likely that visualization tools will not only be used in unexpected ways, but in ways we actively dislike. As new audiences are exposed to visualization technologies, "bad" or "chart junk" visualizations will be generated. Furthermore, visualizations will be used to support actions or points of view we may find distasteful, and any communication medium that is sufficiently powerful to inform may also be used to lie or misrepresent. We as a community should not be so concerned with trying to control the medium or prevent people from lying or creating bad visualizations. As audiences get more comfortable communicating with visualizations, we optimistically expect the quality of visualizations and nuance of interpretation to improve.

However, this proscription does not mean that researchers should idly sit on their hands. Rather, there will be an expanded role for visualization experts to play. Issues of data provenance, cleaning, and integrity will force the research community to focus on the visualization pipeline in a more holistic manner. Supporting data at varied levels of structure will become increasingly necessary. New genres of visualization use may require new designs and new systems to support emerging practices, and the design of visual exploration tools that both empower and educate will take on new importance. Consequently, the entrance of visualization technologies into the mainstream offers a new horizon of research opportunities.

References

1. Agrawala, M., Beers, A.C., McDowall, I., Fröhlich, B., Bolas, M., Hanrahan, P.: The Two-User Responsive Workbench: Support for Collaboration Through Individual Views of a Shared Space. In: International Conference on Computer Graphics and Interactive Techniques (Siggraph '97), pp. 327–332. ACM Press, New York (1997), doi:10.1145/258734.258875
2. Ahlberg, C., Wistrand, E.: IVEE: An information visualization & exploration environment. In: Proceedings of the IEEE Symposium on Information Visualization, Atlanta, GA, October 1995, pp. 66–73. IEEE Computer Society Press, Los Alamitos (1995)
3. Anupam, V., Bajaj, C.L., Schikore, D., Shikore, M.: Representations in distributed cognitive tasks. IEEE Computer 27(7), 37–43 (1994)
4. Artero, A.O., Ferreira de Oliveira, M.C., Levkowitz, H.: Uncovering clusters in crowded parallel coordinates visualizations. In: Proceedings of the IEEE Symposium on Information Visualization (InfoVis), pp. 81–88. IEEE Computer Society Press, Los Alamitos (2004)
5. Baldonado, M.Q.W., Woodruff, A., Kuchinsky, A.: Guidelines for using multiple views in information visualization. In: Proceedings of AVI '00, pp. 110–119. ACM Press, New York (2000), doi:10.1145/345513.345271
6. Benbunan-Fich, R., Hiltz, S.R., Turoff, M.: A comparative content analysis of face-to-face vs. asynchronous group decision making. Decision Support Systems 34(4), 457–469 (2003)
7. Benkler, Y.: Coase's penguin, or, linux and the nature of the firm. Yale Law Journal 112(369) (2002)
8. Benko, H., Wilson, A.D., Baudisch, P.: Precise Selection Techniques for Multi-Touch Screens. In: Proceedings of the Conference on Human Factors in Computing Systems (CHI'06), Montréal, Canada, April 22-27, 2006, pp. 1263–1272. ACM Press, New York (2006), doi:10.1145/1124772.1124963
9. Benko, H., Ishak, E.W., Feiner, S.: Collaborative mixed reality visualization of an archaeological excavation. In: IEEE International Symposium on Mixed and Augmented Reality (ISMAR 2004), Arlington, VA, pp. 132–140 (2004)

10. Bertin, J.: Semiology of Graphics: Diagrams Networks Maps (Translation of: Sémiologie graphique). The University of Wisconsin Press, Madison (1983)
11. Billman, D., Convertino, G., Shrager, J., Pirolli, P., Massar, J.P.: Collaborative intelligence analysis with cache and its effects on information gathering and cognitive bias. In: Human Computer Interaction Consortium Workshop (2006)
12. Brennan, S.E.: How conversation is shaped by visual and spoken evidence. In: Trueswell, Tanenhaus (eds.) Approaches to studying world-situated language use: Bridging the language-as-product and language-as-action traditions, pp. 95–129. MIT Press, Cambridge (2005)
13. Brennan, S.E., Mueller, K., Zelinsky, G., Ramakrishnan, I.V., Warren, D.S., Kaufman, A.: Toward a multi-analyst, collaborative framework for visual analytics. In: IEEE Symposium on Visual Analytics Science and Technology (2006)
14. Brodlie, K.W., Duce, D.A., Gallop, J.R., Walton, J.P.R.B., Wood, J.D.: Distributed and collaborative visualization. Computer Graphics Forum 23(2), 223–251 (2004)
15. Brush, A.J., Bargeron, D., Grudin, J., Gupta, A.: Notification for shared annotation of digital documents. In: Proc. ACM Conference on Human Factors in Computing Systems (CHI'02) (2002)
16. Caillois, R.: Man, Play, and Games. Free Press of Glencoe (1961)
17. Card, S., Mackinlay, J.D., Shneiderman, B.: Readings In Information Visualization: Using Vision To Think. Morgan Kauffman Publishers, Inc., San Francisco (1999)
18. Card, S.K., Nation, D.: Degree-of-Interest Trees: A component of an attention-reactive user interface. In: Proceedings of the Working Conference on Advanced Visual Interfaces, May 2002, pp. 231–245. ACM Press, New York (2002), http://www2.parc.com/istl/projects/uir/pubs/items/UIR-2002-11-Card-AVI-DOITree.pdf
19. Caroll, J., Rosson, M.B., Convertino, G., Ganoe, C.H.: Awareness and teamwork in computer-supported collaborations. Interacting with Computers 18(1), 21–46 (2005)
20. Carpendale, M.S.T., Montagnese, C.: A framework for unifying presentation space. In: Schilit, B. (ed.) Proceedings of ACM Symposium on User Interface Software and Technology (UIST), pp. 61–70. ACM Press, New York (2001), http://pages.cpsc.ucalgary.ca/~sheelagh/personal/pubs/2001/carpendal euist01.pdf, doi:10.1145/502348.502358
21. Carter, S., Mankoff, J., Goddi, P.: Building connections among loosely coupled groups: Hebb's rule at work. Journal of Computer-Supported Cooperative Work 13(3), 305–327 (2004)
22. Cheshire, C.: Selective incentives and generalized information exchange. Social Psychology Quarterly 70(1) (2007)
23. Chi, E.H.-H., Riedl, J.T.: An Operator Interaction Framework for Visualization Systems. In: Wills, G., Dill, J. (eds.) Proceedings of the IEEE Symposium on Information Visualization (InfoVis '98), pp. 63–70. IEEE Computer Society Press, Los Alamitos (1998)
24. Chuah, M.C., Roth, S.F.: Visualizing Common Ground. In: Proc. of the Conf. on Information Visualization (IV), pp. 365–372. IEEE Computer Society Press, Los Alamitos (2003)
25. Chui, Y.-P., Heng, P.-A.: Enhancing view consistency in collaborative medical visu-alization systems using predictive-based attitude estimation. In: First IEEE International Workshop on Medical Imaging and Augmented Reality (MIAR'01), Hong Kong, China (2001)

26. Clark, H.H.: Pointing and placing. In: Kita, S. (ed.) Pointing. Where language, culture, and cognition meet, pp. 243–268. Lawrence Erlbaum, Mahwah (2003)
27. Cluxton, D., Eick, S.G., Yun, J.: Hypothesis visualization. In: Proceedings of the IEEE Symposium on Information Visualization (Posters Compendium), Austin, TX, October 2004, pp. 9–10. IEEE Computer Society Press, Los Alamitos (2004)
28. Dietz, P.H., Leigh, D.L.: Diamondtouch: A multi-user touch technology. In: Proc. ACM Symposium on User Interface Software and Technology, pp. 219–226 (2001)
29. Dourish, P., Belotti, V.: Awareness and coordination in shared workspaces. In: Proc. ACM Conference on Computer-Supported Cooperative Work, Toronto, Ontario, pp. 107–114 (1992)
30. Dourish, P., Chalmers, M.: Running out of space: Models of information navigation. In: Proc. Human Computer Interaction (HCI'94) (1994)
31. Dynamics, G.: Command post of the future. Website (accessed November 2007)
32. Eccles, R., Kapler, T., Harper, R., Wright, W.: Stories in geotime. In: Proc. IEEE Symposium on Visual Analytics Science and Technology (2007)
33. Ellson, J., Gansner, E.R., Koutsofios, E., North, S.C., Woodhull, G.: Graphviz and Dynagraph – static and dynamic graph drawing tools. Online Documentation (accessed November 2007)
34. Fekete, J.-D.: The Infovis Toolkit. In: Ward, M., Munzner, T. (eds.) Proceedings of the IEEE Symposium on Information Visualization (InfoVis), pp. 167–174. IEEE Computer Society Press, Los Alamitos (2004)
35. Forlines, C., Shen, C.: DTLens: Multi-user Tabletop Spatial Data Exploration. In: Proc. of User Interface Software and Technology (UIST), pp. 119–122. ACM Press, New York (2005), doi:10.1145/1095034.1095055
36. Gahegan, M., Wachowicz, M., Harrower, M., Rhyne, T.-M.: The integration of geographic visualization with knowledge discovery in databases and geocomputation. Cartography and Geographic Information Society 28(1), 29–44 (2001)
37. Gershon, N.: What storytelling can do for information visualization. Communications of the ACM 44(8), 31–37 (2001)
38. Grimstead, I.J., Walker, D.W., Avis, N.J.: Collaborative visualization: A review and taxonomy. In: Proceedings of the Symposium on Distributed Simulation and Real-Time Applications, pp. 61–69. IEEE Computer Society Press, Los Alamitos (2005)
39. Gutwin, C., Greenberg, S.: Design for individuals, design for groups: Tradeoffs between power and workspace awareness. In: Proceedings of Computer Supported Cooperative Work (CSCW), pp. 207–216. ACM Press, New York (1998), doi:10.1145/289444.289495
40. Hancock, M., Carpendale, S.: Supporting multiple off-axis viewpoints at a tabletop display. In: Proceedings of Tabletop, pp. 171–178. IEEE Computer Society Press, Los Alamitos (2007)
41. Heer, J.: The flare visualization toolkit. Website (accessed November 2007)
42. Heer, J.: Socializing visualization. In: Proc. CHI 2006 Workshop on Social Visualization (2006)
43. Heer, J., Agrawala, M.: Design Considerations for Collaborative Visual Analytics. In: IEEE Symposium on Visual Analytics Science and Technology (VAST), pp. 171–178. IEEE Computer Society Press, Los Alamitos (2007), http://vis.berkeley.edu/papers/design_collab_vis/2007-DesignCollabVis-VAST.pdf
44. Heer, J., Card, S.K., Landay, J.A.: prefuse: A toolkit for interactive information visualization. In: Proceedings of the Conference on Human Factors in Computing Systems (CHI), pp. 421–430. ACM Press, New York (2005), doi:10.1145/1054972.1055031

45. Heer, J., boyd, d.: Vizster: Visualizing Online Social Networks. In: Proceedings of the IEEE Symposium on Information Visualization (InfoVis '05), pp. 33–40. IEEE Computer Society Press, Los Alamitos (2005)
46. Heer, J., Viégas, F.B., Wattenberg, M.: Voyagers and voyeurs: Supporting asynchronous collaborative information visualization. In: Proceedings of the Conference on Human Factors in Computing Systems (CHI), pp. 1029–1038. ACM Press, New York (2007)
47. Hill, W.C., Hollan, J.D.: Deixis and the future of visualization excellence. In: Proc. of IEEE Visualization, pp. 314–319 (1991)
48. Isenberg, T., Neumann, P., Carpendale, S., Nix, S., Greenberg, S.: Interactive annotations on large, high-resolution information displays. In: Conference Compendium of IEEE VIS, InfoVis, and VAST, pp. 124–125. IEEE Computer Society Press, Los Alamitos (2006), http://cpsc.ucalgary.ca/~isenberg/papers/Isenberg_2006_IAL.pdf
49. Jerding, D.F., Stasko, J.T.: The Information Mural: A Technique for Displaying and Navigating Large Information Spaces. In: Proceedings of the IEEE Symposium on Information Visualization (InfoVis), pp. 43–50. IEEE Computer Society Press, Los Alamitos (1995)
50. Johnson, B., Shneiderman, B.: Tree-maps: A space-filling approach to the visualization of hierarchical information structures. In: Proceedings of IEEE Visualization, pp. 284–291. IEEE Computer Society Press, Los Alamitos (1991)
51. Kerren, A., Stasko, J.T., Fekete, J.-D., North, C.J. (eds.): Information Visualization. LNCS, vol. 4950. Springer, Heidelberg (2008)
52. Kleinmutz, D.N., Schkade, D.A.: Information Displays and Decision Processes. Psychological Science 4(4), 221–227 (1993)
53. Kruger, R., Carpendale, S., Scott, S.D., Greenberg, S.: Roles of orientation in tabletop collaboration: Comprehension, coordination and communication. Journal of Computer Supported Collaborative Work 13(5–6), 501–537 (2004)
54. Kruger, R., Carpendale, S., Scott, S.D., Tang, A.: Fluid integration of rotation and translation. In: Proceedings of Human Factors in Computing Systems (CHI), pp. 601–610. ACM Press, New York (2005), doi:10.1145/1054972.1055055
55. Lamping, J., Rao, R., Pirolli, P.: A focus + context technique based on hyperbolic geometry for visualizing large hierarchies. In: Proceedings of the Conference of Human Factors in Computing Systems,CHI, pp. 401–408. ACM Press, New York (1995), doi:10.1145/223904.223956
56. Ling, K., Beenen, G., Lundford, P., Wang, X., Chang, K., Li, X., Cosley, D., Frankowski, D., Terveen, L., Rashid, A.M., Resnick, P., Kraut, R.: Using social psychology to motivate contributions to online communities. Journal of Computer-Mediated Communication 10(4) (2005)
57. Liu, Y., Gahegan, M., Macgill, J.: Increasing geocomputational interoperability: Towards a standard geocomputation API. In: Proceedings of GeoComputation, Ann Arbor, MI (2005)
58. Livny, M., Ramakrishnan, R., Beyer, K., Chen, G., Donjerkovic, D., Lawande, S., Myllymaki, J., Wenger, K.: DEVise: Integrated querying and visualization of large datasets. In: Proceedings of the International Conference on Management of Data (SIGMOD), Tucson, AZ, pp. 301–312. ACM Press, New York (1997)
59. Mackinlay, J.D.: Automating the design of graphical presentations of relational information. ACM Transactions on Graphics 5(2), 110–141 (1986)
60. Mackinlay, J.D., Hanrahan, P., Stolte, C.: Show me: Automatic presentation for visual analysis. IEEE Transactions on Visualization and Computer Graphics 13(6), 1137–1144 (2007)

61. Mark, G., Kobsa, A.: The effects of collaboration and system transparency on CIVE usage: An empirical study and model. Presence 14(1), 60–80 (2005)
62. Marr, D.: Vision: A Computational Investigation into the Human Representation and Processing of Visual Information. W.H. Freeman, New York (1982)
63. Merriam-Webster: Webster's english dictionary. Website (accessed November 2007), http://www.cs.chalmers.se/~hallgren/wget.cgi?presentation
64. Munzner, T.: Guest Editor's Introduction: Information Visualization. Computer Graphics and Applications 22(1), 20–21 (2002)
65. Munzner, T., Guimbretiére, F., Tasiran, S., Zhang, L., Zhou, Y.: TreeJuxtaposer: Scalable Tree Comparison Using Focus+Context with Guaranteed Visibility. ACM Transactions on Graphics 22(3), 453–462 (2003)
66. Nacenta, M.A., Sakurai, S., Yamaguchi, T., Miki, Y., Itoh, Y., Kitamura, Y., Subramanian, S., Gutwin, C.: E-conic: a perspective-aware interface for multi-display environments. In: Proceedings of ACM Symposium on User Interface Software and Technology (UIST'01), pp. 279–288. ACM Press, New York (2007)
67. Neumann, P., Schlechtweg, S., Carpendale, M.S.T.: ArcTrees: Visualizing Relations in Hierarchical Data. In: Proceedings of Eurographics / IEEE VGTC Symposium on Visualization (EuroVis 2005, June 1–3, 2005, Leeds, England, UK), Aire-la-Ville. Eurographics Workshop Series, pp. 53–60. Eurographics (2005)
68. Neumann, P., Tang, A., Carpendale, S.: A Framework for Visual Information Analysis. Technical Report 2007-87123, University of Calgary, Calgary, AB, Canada (July 2007)
69. North, C., Shneiderman, B.: Snap-together visualization: A user interface for coordinating visualizations via relational schemata. In: Proceedings of the Working Conference on Advanced Visual Interfaces (AVI), May 2000, pp. 128–135. ACM Press, New York (2000)
70. Olson, G.M., Olson, J.S.: Distance Matters. Human-Computer Interaction 15(2 & 3), 139–178 (2000)
71. Olston, C., Woodruff, A., Aiken, A., Chu, M., Ercegovac, V., Lin, M., Spalding, M., Stonebraker, M.: Datasplash. In: Proceedings of the International Conference on Management of Data (SIGMOD), Seattle, WA, June 1998, pp. 550–552. ACM Press, New York (1998)
72. Park, K.S., Kapoor, A., Leigh, J.: Lessons learned from employing multiple perspectives in a collaborative virtual environment for visualizing scientific data. In: Proceedings of Collaborative Virtual Environments (CVE), pp. 73–82. ACM Press, New York (2000), doi:10.1145/351006.351015
73. Pinelle, D., Gutwin, C., Greenberg, S.: Task analysis for groupware usability evaluation: Modeling shared-workspace tasks with the mechanics of collaboration. ACM Transaction of Human Computer Interaction 10(4), 281–311 (2003)
74. Rensink, R.A.: chapter Change Blindness. In: McGraw-Hill Yearbook of Science and Technology, pp. 44–46. McGraw-Hill, New York (2005)
75. Meredith, Ryall, K., Shen, C., Forlines, C., Morris, F.V.R.: Beyond "social protocols": Multi-user coordination policies for co-located groupware. In: Proceedings of Computer-Supported Cooperative Work (CSCW), pp. 262–265. ACM Press, New York (2004), doi:10.1145/1031607.1031648
76. Rogers, Y., Lindley, S.: Collaborating around vertical and horizontal large interactive displays: Which way is best? Interacting with Computers 16(6), 1133–1152 (2004)
77. Meredith, Everitt, K., Ryall, F.V.K., Esenther, A., Forlines, C., Shen, C., Shipman, S., Morris, R.: Identity-differentiating widgets for multiuser interactive surfaces. IEEE Computer Graphics and Applications 26(5), 56–64 (2006)

78. Salen, K., Zimmerman, E.: Rules of Play: Fundamentals of Game Design. MIT Press, Cambridge (2003)
79. Saraiya, P., North, C., Duca, K.: An Insight-Based Methodology for Evaluating Bioinformatics Visualizations. IEEE Transactions on Visualization and Computer Graphics 11(4), 443–456 (2005)
80. Scott, S.D., Carpendale, M.S.T., Habelski, S.: Storage bins: Mobile storage for collaborative tabletop displays. IEEE Computer Graphics and Applications 25(4), 58–65 (2005), `http://doi.ieeecomputersociety.org/10.1109/MCG.2005.86`
81. Scott, S.D., Carpendale, M.S.T., Inkpen, K.M.: Territoriality in collaborative tabletop workspaces. In: Proceedings of Computer-Supported Cooperative Work (CSCW), pp. 294–303. ACM Press, New York (2004), `http://innovis.cpsc.ucalgary.ca/pubs/2004/Territoriality.CSCW/scott_cscw2004.pdf`, doi:10.1145/1031607.1031655
82. Scott, S.D., Grant, K.D., Mandryk, R.L.: System guidelines for co-located collaborative work on a tabletop display. In: Proceedings of the European Conference on Computer-Supported Cooperative Work (ECSCW), pp. 159–178. Kluwer Academic Publishers, Dordrecht (2003), `http://www.ecscw.uni-siegen.de/2003/009Scott_ecscw03.pdf`
83. Seo, J., Shneiderman, B.: Knowledge discovery in high dimensional data: Case studies and a user survey for an information visualization tool. IEEE Transactions on Visualization and Computer Graphics 12(3), 311–322 (2006)
84. Shen, C., Lesh, N., Vernier, F.: Personal digital historian: Story sharing around the table. ACM Interactions 10(2), 15–22 (2003)
85. Shneiderman, B.: The eyes have it: A task by data type taxonomy for information visualizations. In: Proceedings of the IEEE Symposium on Visual Languages, pp. 336–343. IEEE Computer Society Press, Los Alamitos (1996)
86. Shum, S.B., Li, V.U.G., Domingue, J., Motta, E.: Visualizing internetworked argumentation. In: Kirschner, P.A., Shum, S.J.B., Carr, C.S. (eds.) Visualizing Argumentation: Software Tools for Collaborative and Educational Sense-Making, December 2002, pp. 185–204. Springer, Heidelberg (2002)
87. Spence, R.: Information Visualization, 2nd edn. Pearson Education Limited, Harlow (2007)
88. Takatsuka, M., Gahegan, M.: GeoVISTA Studio: A codeless visual programming environment for geoscientific data analysis and visualization. Computational Geoscience 28(10), 1131–1144 (2002)
89. Tandler, P., Prante, T., Müller-Tomfelde, C., Streitz, B., Steinmetz, R.: ConnecTables: Dynamic Coupling of Displays for the Flexible Creation of Shared Workspaces. In: Proceedings of User Interface Software and Technology (UIST), pp. 11–20. ACM Press, New York (2001)
90. Tang, A., Tory, M., Po, B., Neumann, P., Carpendale, S.: Collaborative coupling over tabletop displays. In: Proceedings of Human Factors in Computing Systems (CHI), pp. 1181–1290. ACM Press, New York (2006), doi:10.1145/1124772.1124950
91. Thomas, J.J., Cook, K.A.: Illuminating the Path: The Research and Development Agenda for Visual Analytics. National Visualization and Analytics Center (2005), `http://nvac.pnl.gov/agenda.stm`
92. Tufte, E.R.: The Visual Display of Quantitative Information. Graphic Press, Cheshire (2001)
93. van Wijk, J.J.: The value of visualization. In: Proceedings of IEEE Visualization (VIS), pp. 79–86. IEEE Computer Society Press, Los Alamitos (2005), `http://www.win.tue.nl/~vanwijk/vov.pdf`

94. Vernier, F., Lesh, N., Shen, C.: Visualization techniques for circular tabletop interfaces. In: Proceedings of Advanced Visual Interfaces (AVI), pp. 257–263. ACM Press, New York (2002)
95. Viégas, F.B., boyd, d., Nguyen, D.H., Potter, J., Donath, J.: Digital artifacts for remembering and storytelling: PostHistory and social network fragments. In: Proceedings of the Hawaii International Conference on System Sciences (HICCSS), pp. 105–111 (2004)
96. Viégas, F.B., Wattenberg, M.: Communication-minded visualization: A call to action. IBM Systems Journal 45(4), 801–812 (2006), doi:10.1147/sj.454.0801
97. Viégas, F.B., Wattenberg, M., van Ham, F., Kriss, J., McKeon, M.: Many Eyes: A site for visualization at internet scale. IEEE Transactions on Visualization and Computer Graphics (Proceedings Visualization / Information Visualization 2007) 12(5), 1121–1128 (2007),
    http://www.research.ibm.com/visual/papers/viegasinfovis07.pdf
98. von Ahn, L.: Games with a purpose. Computer 39(6), 92–94 (2006)
99. Ware, C.: Information Visualization – Perception for Design, 2nd edn. Morgan Kaufmann Series in Interactive Technologies. Morgan Kaufmann Publishers, San Francisco (2004)
100. Wattenberg, M., Kriss, J.: Designing for Social Data Analysis. IEEE Transactions on Visualization and Computer Graphics 12(4), 549–557 (2006)
101. Weaver, C., Fyfe, D., Robinson, A., Holdsworth, D.W., Peuquet, D.J., MacEachren, A.M.: Visual analysis of historic hotel visitation patterns. In: Proceedings of the Symposium on Visual Analytics Science and Technology (VAST), Baltimore, MD, October 31–November 2 2006, pp. 35–42. IEEE Computer Society Press, Los Alamitos (2006)
102. Weaver, C.E.: Improvise: A User Interface for Interactive Construction of Highly-Coordinated Visualizations. Phd thesis, University of Wisconsin–Madison (June 2006)
103. Wesche, G., Wind, J., Göbe, M., Rosenblum, L., Durbin, J., Doyle, R., Tate, D., King, R., Fröhlich, B., Fischer, M., Agrawala, M., Beers, A., Hanrahan, P., Bryson, S.: Application of the Responsive Workbench. IEEE Computer Graphics and Applications 17(4), 10–15 (1997), doi:10.1109/38.595260
104. Weskamp, M.: newsmap. Website (accessed November 2007),
    http://marumushi.com/apps/newsmap/index.cfm
105. Wigdor, D., Shen, C., Forlines, C., Balakrishnan, R.: Perception of elementary graphical elements in tabletop and multi-surface environments. In: Proceedings of Human Factors in Computing Systems (CHI), pp. 473–482. ACM Press, New York (2007)
106. Willett, W., Heer, J., Agrawala, M.: Scented widgets: Improving navigation cues with embedded visualizations. IEEE Transactions on Visualization and Computer Graphics 13(6), 1129–1136 (2007)
107. Wright, W., Schroh, D., Proulx, P., Skaburskis, A., Cort, B.: Advances in nSpace – the sandbox for analysis. In: International Conference on Intelligence Analysis, McLean, VA (May 2005)
108. Yang, D., Rundensteiner, E.A., Ward, M.O.: Analysis guided visual exploration to multivariate data. In: Proc. IEEE Visual Analytics Science and Technology (2007)
109. Yost, B., North, C.: The perceptual scalability of visualization. IEEE Transactions on Visualization and Computer Graphics 12(5), 837–844 (2005)
110. Zhang, J., Norman, D.A.: Representations in distributed cognitive tasks. Cognitive Science 18(1), 87–122 (1994)

111. Zuk, T., Schlesier, L., Neumann, P., Hancock, M.S., Carpendale, M.S.T.: Heuristics for Information Visualization Evaluation. In: Proceedings of the Workshop Beyond Time and Errors (BELIV), held in conjunction with AVI, pp. 55–60. ACM Press, New York (2006), doi:10.1145/1168149.1168162

Process and Pitfalls in Writing Information Visualization Research Papers

Tamara Munzner

Department of Computer Science,University of British Columbia
201-2366 Main Mall, Vancouver BC V6T 1Z4 Canada,
tmm@cs.ubc.ca,
http://www.cs.ubc.ca/~tmm

Abstract. The goal of this paper is to help authors recognize and avoid a set of pitfalls that recur in many rejected information visualization papers, using a chronological model of the research process. Selecting a target paper type in the initial stage can avert an inappropriate choice of validation methods. Pitfalls involving the design of a visual encoding may occur during the middle stages of a project. In a later stage when the bulk of the research is finished and the paper writeup begins, the possible pitfalls are strategic choices for the content and structure of the paper as a whole, tactical problems localized to specific sections, and unconvincing ways to present the results. Final-stage pitfalls of writing style can be checked after a full paper draft exists, and the last set of problems pertain to submission.

1 Introduction

Many rejected information visualization research papers have similar flaws. In this paper, I categorize these common pitfalls in the context of stages of the research process. My main goal is to help authors escape these pitfalls, especially graduate students or those new to the field of information visualization. Reviewers might also find these pitfalls an interesting point of departure when considering the merits of a paper.

This paper is structured around a chronological model of the information visualization research process. I argue that a project should begin with a careful consideration of the type of paper that is the desired outcome, in order to avoid the pitfalls of unconvincing validation approaches. Research projects that involve the design of a new visual encoding would benefit from checking for several middle-stage pitfalls in unjustified or inappropriate encoding choices. Another critical checkpoint is the late stage of the project, after the bulk of the work is done, but before diving in to writing up results. At this point, you should consider both strategic pitfalls about the high-level structure of the entire paper, tactical pitfalls that affect one or a few sections, and possible pitfalls in the specifics of your approach to the results section. At a final stage, when there is a complete paper draft, you can check for lower-level pitfalls of writing style, and avoid submission-time pitfalls.

A. Kerren et al. (Eds.): Information Visualization, LNCS 4950, pp. 134–153, 2008.

I have chosen a breezy style, following in the footsteps of Levin and Re-dell [22] and Shewchuk [34]. My intent is serious, but I have tried to invent catchy – sometimes even snide – titles in hopes of making these pitfalls more memorable. Guides to writing research papers have been written in several sub-fields of computer science, including systems [22], software engineering [33], programming languages [19], networking [28], and graphics [20]. Many of the pitfalls in the middle and later project stages apply to research writing in general, not just information visualization, and have been mentioned in one or many of these previous papers.

My first pass at providing advice for authors and reviewers in the field of information visualization, abbreviated as **infovis**, was the creation of the author guide for the annual conference. When I was Posters Chair of InfoVis 2002, the IEEE Symposium on Information Visualization, I read the roughly 300 reviews of the 78 rejected papers in order to decide which to invite as poster submissions. The experience convinced me that future paper authors would benefit from more specific guidance. When I became Papers Chair in 2003, with co-chair Stephen North, we completely rewrote the Call for Papers. We introduced five categories of papers, with an explicit discussion of the expectations for each, in a guide for authors that has been kept unchanged through the 2007 conference.

This second pass is motivated by the patterns of mistakes I saw in my two-year term as InfoVis Papers Co-Chair where I read the over 700 reviews for all 189 submitted papers, and in personally writing nearly 100 reviews in the subsequent three years. My discussion of paper types below expands considerably on the previous author guide, and I provide concrete examples of strong papers for each type. The advice I offer is neither complete nor objective; although I draw on my experience as a papers chair, my conclusions may be idiosyncratic and reflect my personal biases. I do not perform any quantitative analysis. Doing so in the domain of infovis would, no doubt, be fruitful future work, given the interesting results from software engineering [33] and human-computer interaction [17].

None of these pitfalls are aimed at any particular individual: I have seen multiple instances of each one. Often a single major pitfall was enough to doom a paper to rejection, although in some cases I have seen other strengths outweigh a particular weakness. I hasten to point out that I, myself, have committed some of the errors listed below, and despite my best efforts I may well fall prey to them in the future.

2 Initial Stage: Paper Types

A good way to begin a research project is to consider where you want it to end. That is, what kind of a paper do you intend to produce? That choice should guide many of your downstream decisions, including the critical issue of how to validate any claims you might make about your research contribution.

2.1 Validation Approaches

Many possible ways exist to validate the claim that an infovis paper has made a contribution, including:

- algorithm complexity analysis
- implementation performance (speed, memory)
- quantitative metrics
- qualitative discussion of result pictures
- user anecdotes (insights found)
- user community size (adoption)
- informal usability study
- laboratory user study
- field study with target user population
- design justification from task analysis
- visual encoding justification from theoretical principles

In any particular paper, the constraints of researcher time and page limits force authors to select a subset of these approaches to validation. The taxonomy of paper types below can provide you with considerable guidance in choosing appropriate validation approaches, leading to a paper structure where your results back up your claims. The five paper types guide the presentation of your research by distinguishing between the following possibilities for your primary contribution: an algorithm, a design, a system, a user study, or a model.

2.2   Technique

Technique papers focus on novel algorithms and an implementation is expected. The most straightforward case is where the research contribution is a new algorithm that refines or improves a technique proposed in previous work. A typical claim is that the new algorithm is faster, more scalable, or provides better visual quality than the previously proposed one. The MillionVis system [5], hierarchical parallel coordinates [6], and hierarchical edge bundling [15] are good exemplars for this category.

Typical results to back up such a claim would be algorithm complexity analysis, quantitative timing measurements of the implementation, and a qualitative discussion of images created by the new algorithm. Quantitative metrics of image quality, for example edge crossings in graph layout, are also appropriate. You need to compare these results side by side against those from competing algorithms. You might collect this information through some combination of using results from previous publications, running publicly available code, or implementing them yourself. In this case, there is very little or no design justification for whether the technique is actually suitable for the proposed problem domain in the paper itself: there is an implicit assumption that the previous cited work makes such arguments.

In retrospect, a better name for this category might be **Algorithms**. Many authors who design new visual representations might think that a paper documenting a new technique belongs in the Technique category. However, the question to ask is whether your primary contribution is the algorithm itself, or the design. If your algorithm is sophisticated enough that it requires several pages of description for replicability, then you probably have a primary algorithmic contribution. If the algorithm itself is straightforward enough that only a brief description is required, or if all of the techniques that you use have been adequately described in previous work, then you would be better served by explicitly writing a Design Study paper.

2.3   Design Study

Design Study papers make a case that a new visual representation is a suitable solution for a particular domain problem. First, you should explain the target problem. You must provide enough background that the reader can pass judgement about whether your solution is good, but not so much detail that the focus of the paper is on domain problems rather than infovis issues. Finding the right balance is a difficult but crucial judgement call. Second, you should crisply state the design requirements that you have determined through your task analysis. Third, you should present your visual encoding and interaction mechanisms and justify these design choices in terms of how well they fulfill the requirements. Typical arguments would refer to perceptual principles and infovis theory. For example, using spatial position to encode the most important variables and using greyscale value rather than hue to encode an ordered variable are both very defensible choices [24]. The best justifications explicitly discuss particular choices in the context of several possible alternatives.

Fourth, you should present results that back up the claim that your approach is better than others. Typical results include case studies or scenarios of use. Design studies often document iterative design and the use of formative evaluation for refinement. The research contribution of a design study is not typically a new algorithm or technique, but rather a well-reasoned justification of how existing techniques can be usefully combined. For most design studies, adoption by the target users is valuable evidence that the system has met its goals, as are anecdotes of insights found with the new system that would be difficult to obtain using previous methods.

I think this category name is still a good choice, despite the fact that great design studies are all too rare. I argue that the field would be well served if more authors explicitly cast their work in this category. Interesting examples that approach the design study from several angles are the cluster-calendar system [40], ThemeRiver [11], Vizster [13], VistaChrom [21], and a hotel visitation pattern analysis system [42].

2.4   Systems

Systems papers focus on the architectural choices made in the design of an infrastructure, framework, or toolkit. A systems paper typically does not introduce new techniques or algorithms. A systems paper also does not introduce a new design for an application that solves a specific problem; that would be a design study. The research contribution of a systems paper is the discussion of architectural design choices and abstractions in a framework or library, not just a single application. A good example is the prefuse systems paper [14], which has a discussion of the performance, flexibility, and extensibility implications of the ItemRegistry, Action, and ActionList data structure choices. Another good example is the systems paper on design choices made in Rivet [37] and other systems with similar goals, such as the tradeoffs of data granularity for transformations.

A systems paper can be considered as a specialized kind of design study: one about the choices made when building a library as opposed to the choices made when solving a visual encoding problem. Like the design study category, key aspects of a systems paper are the lessons learned from building the system, and observing its use. I urge authors and reviewers of systems papers to peruse Levin and Redell's classic on "How (and How Not) to Write a Good Systems Paper" [22].

The category name might be a cause for confusion because the the term system is often used interchangeably with application or implementation. The original intent was to follow the distributed systems usage where there is a very strong distinction between system-level and application-level work. Although a name like **Toolkit** might avert that confusion, the term 'systems paper' is such a strong convention in computer science that I am reluctant to advocate this change.

2.5   Evaluation

Evaluation papers focus on assessing how an infovis system or technique is used by some target population. Evaluation papers typically do not introduce new techniques or algorithms, and often use implementations described in previous work. The most common approach in infovis thus far has been formal user studies conducted in laboratory setting, using carefully abstracted tasks that can be quantitatively measured in terms of time and accuracy, and analyzed with statistical methods. A typical claim would be that the tested tasks are ecologically valid; that is, they correspond to those actually undertaken by target users in a target domain. A typical result would be a statistically significant main effect of an experimental factor, or interaction effect between factors. The work of Yost and North on perceptual scalability is a good example of this subtype [44]. A different approach to studying user behavior is field studies, where a system is deployed in a real-world setting with its target users. In these studies, the number of participants is usually smaller, with no attempt to achieve statistical significance, and the time span is usually weeks or months rather than hours. However, the study design axes of field versus laboratory, short-term versus long-term, and size are all orthogonal. Both quantitative and qualitative measurements may be collected. For example, usage patterns may be studied through quantitative logging of mouse actions or eyegaze. The work of Hornbæk and Hertzum on untangling fisheye menus is a good example of this subtype [16]. Usage patterns can also be studied through qualitative observations during the test itself or later via coding of videotaped sessions. Trafton *et al.*'s field study of how meteorologists use visual representations is an excellent example of the power of video coding [39].

Plaisant's discussion of the difficulties of evaluating infovis is thoughtful and germane [29]. In retrospect, a better, albeit wordy, name for this category might be **Summative User Studies**, since the goal is to examine the strengths of a system or technique. Evaluation is far too broad a term – because all papers should contain some sort of validation. Even User Studies would not be the best choice, because formative studies are probably a better fit for the Design Study category, where ethnographic methods are often appropriate in the task analysis to determine design requirements or iteratively refine a design. However, these lines are not necessarily crisp. For instance, the MILC approach advocated by Shneiderman and Plaisant [35] could fit into either a Design Study framework, if the emphasis is on the formative ethnographic analysis and iterative design, or a Summative Evaluation framework, if the emphasis is on the longitudinal field study.

2.6   Model

Model papers present formalisms and abstractions as opposed to the design or evaluation of any particular technique or system. This category is for meta-research papers, where the broad purpose is to help other researchers think about their own work.

The most common subcategory is **Taxonomy**, where the goal is to propose categories that help researchers better understand the structure of the space of possibilities for some topic. Some boundaries will inevitably be fuzzy, but the goal is to be as comprehensive and complete as possible. As opposed to a survey paper, where the goal is simply to summarize the previous work, a taxonomy paper proposes some new categorization or expands upon a previous one and may presume the reader's familiarity with the previous work. Good examples are Card and Mackinlay's taxonomy of visual encodings [3] and Amar *et al.*'s task taxonomy [1].

A second subcategory is **Formalism**, for papers that present new models, definitions, or terminology to describe techniques or phenomena. A key attribute of these kinds of papers is reflective observation. The authors look at what is going on in a field and provide a new way of thinking about it that is clear, insightful, and summative. An influential example is the space-scale diagram work of Furnas and Bederson [7], and an interesting recent example is the casual infovis definition from Pousman *et al.* [31].

A third subcategory is **Commentary**, where the authors advocate a position and argue to support it. Typical arguments would be "the field needs to do more X", "we should be pushing for more Y", or "avoid doing Z because of these drawbacks". A good example is the fisheye followup from Furnas [8]. These kinds of papers often cite many examples and may also introduce new terminology.

Model papers can provide both a valuable summary of a topic and a vocabulary to more concisely discuss concepts in the area. They can be valuable for both established researchers and newcomers to a field, and are often used as assigned readings in courses. I think this category name is appropriate and do not suggest changing it.

2.7 Combinations

These categories are not hard and fast: some papers are a mixture. For example, a design study where the primary contribution is the design might include a secondary contribution of summative evaluation in the form of a lab or field study. Similarly, a design study may have a secondary contribution in the form of a novel algorithm. Conversely, a technique paper where the primary contribution is a novel algorithm may also include a secondary design contribution in the form of a task analysis or design requirements. However, beware the *Neither Fish Nor Fowl* pitfall I discuss below.

2.8 Type Pitfalls

Carefully consider the primary contribution of your work to avoid the pitfalls that arise from a mismatch between the strengths of your project and the paper type you choose.

Design in Technique's Clothing: Don't validate a new design by providing only performance measurements. Many rejected infovis papers are bad design studies, where a new design is proposed but the design requirements are not crisply stated and justification for design choices is not presented. Many of these authors would be surprised to hear that they have written a design study, because they simply assume that the form of a technique paper is always the correct choice. Technique papers are typically appropriate if you have a novel algorithmic contribution, or you are the very first to propose a technique. If you are combining techniques that have been previously proposed, then the design study form is probably more appropriate.

Application Bingo versus Design Study: Don't apply some random technique to a new problem without thoroughly thinking about what the problem is, whether the technique is suitable, and to what extent it solves the problem. I define 'application bingo' as the game where you pick a narrowly defined problem domain, a random technique, and then write an application paper with the claim of novelty for this particular domain-technique combination. Application bingo is a bad game to play because an overwhelming number of the many combinatorial possibilities lead to a bad design.

Although application bingo is admittedly a caricature, the important question is how we can distinguish those who inadvertently play it from those who genuinely solve a domain problem with an effective visualization. Some visualization venues distinguish between research papers and what are called applications or case studies. This paper category is often implicitly or explicitly considered to be a way to gather data from a community outside of visualization itself. Although that goal is laudable, the mechanism has dangers. A very common pitfall is that application paper submissions simply describe an instantiation of a previous technique in great detail. Many do not have an adequate description of the domain problem. Most do not have an adequate justification of why the technique is suitable for the problem. Most do not close the loop with a validation that the proposed solution is effective for the target users.

In contrast, a strong design study would be rather difficult for an outsider unfamiliar with the infovis literature to write. Two critical aspects require a thorough understanding of the strengths and weaknesses of many visualization techniques. First, although a guideline like "clearly state the problem" might seem straightforward at first glance, the job of abstracting from a problem in some target domain to design requirements *that can be addressed through visualization techniques* requires knowing those techniques. Second, justifying why the chosen techniques are more appropriate than other techniques again requires knowledge of the array of possible techniques.

The flip side of this situation is that design studies where visualization researchers do not have close contact with the target users are usually also weak. A good methodology is collaboration between visualization researchers and target users with driving problems [18](Chapter 3.4).

All That Coding Means I Deserve a Systems Paper: Many significant coding efforts do not lead to a systems paper. Consider whether or not you have specific architectural lessons to offer to the research community that you learned as a result of building your library or toolkit.

Neither Fish nor Fowl: Papers that try to straddle multiple categories often fail to succeed in any of them. Be ruthlessly clear about identifying your most important contribution as primary, and explicitly categorize any other contributions as secondary. Then make structural and validation choices based on the category of the single primary contribution.

3   Middle Pitfalls: Visual Encoding

If you have chosen the design route, then a major concern in the middle stages of a project should be whether your visual encoding choices are appropriate and justifiable.

Unjustified Visual Encoding: An infovis design study paper must carefully justify why the visual encoding chosen is appropriate for the problem at hand. In the case of technique papers, where the focus is on accelerating or improving a previously proposed technique, the argument can be extremely terse and use a citation to a previous paper. But in the case of a design study, or a paper proposing a completely new technique, your justification needs to be explicit and convincing. One of the most central challenges in information visualization is designing the visual encoding and interaction mechanisms to show and manipulate a dataset.

A straightforward visual encoding of the exact input data is often not sufficient. In many successful infovis approaches, the input data undergoes significant transformations into some derived model that is ultimately shown. Many weak papers completely skip the step of task analysis. Without any discussion of the design requirements, it is very hard to convince a reader that your model will solve the problem. In particular, you should consider how to make the case that the structure you are visually showing actually benefits the target end user. For example, many authors new to information visualization simply assert, without justification, that showing the hyperlink structure of the web will benefit end users who are searching for information. One of my own early papers fell prey to this very pitfall [26]. However, after a more careful task analysis, I concluded that most searchers do not need to build a mental model of the structure of the search space, so showing them that structure adds cognitive load rather than reduces it. In a later paper [25], I argued that a visual representation of that hyperlink structure could indeed benefit a specific target community, that of webmasters and content creators responsible for a particular site.

The foundation of information visualization is the characterization of how known facts about human perception should guide visual encoding of abstract datasets. The effectiveness of perceptual channels such as spatial position, color, size, shape, and so on depends on whether the data to encode is categorical, ordered, or quantitative [24]. Many individual perceptual channels are preattentively processed in parallel, yet most combinations of these channels must be serially searched [12]. Some perceptual channels are easily separable, but other combinations are not [41, Chapter 5]. These principles, and many others, are a critical part of infovis theory. The last three pitfalls in this section are a few particularly egregious examples of ignoring this body of knowledge.

Hammer in Search of Nail: If you simply propose a nifty new technique with no discussion of who might ever need it, it's difficult to judge its worth. I am not arguing that all new techniques need to be motivated by specific domain problems: infovis research that begins from a technique-driven starting place can be interesting and stimulating. Moreover, it may be necessary to build an interactive prototype and use it for dataset exploration before it's possible to understand the capabilities of a proposed technique.

However, before you write up the paper about that hammer, I urge you to construct an understanding what kind of nails it can handle. Characterize, at least with some high-level arguments, the kinds of problems where your new technique shines as opposed to those where it performs poorly.

2D Good, 3D Better: The use of 3D rather than 2D for the spatial layout of an abstract dataset requires careful justification that the benefits outweigh the costs [36]. The use of 3D is easy to justify when a meaningful 3D representation is implicit in the dataset, as in airflow over an airplane wing in flow visualization or skeletal structure in medical visualization. The benefit of providing the familiar view is clear, because it matches the mental model of the user. However, when the spatial layout is chosen rather than given, as in the abstract datasets addressed through infovis, there is an explicit choice about which variables to map to spatial position. It is unacceptable, but all too common with naive approaches to infovis, to simply assert that using an extra dimension must be a good idea.

The most serious problem with a 3D layout is occlusion. The ability to interactively change the point of view with navigational controls does not solve the problem. Because of the limitations of human memory, comparing something visible with memories of what was seen before is more difficult than comparing things simultaneously visible side by side [30]. A great deal of work has been devoted to the exploration of the power of multiple linked 2D views, often directly showing derived variables [43] or using them for ordering [2,23]. In many cases, these views are more effective than simply jumping to 3D [40]. Other difficulties of 3D layouts are that users have difficulty in making precise length judgements because of perspective foreshortening [38], and also that text is difficult to read unless it is billboarded to a 2D plane [9].

Color Cacophony: An infovis paper loses credibility when you make design decisions with blatant disregard for basic color perception facts. Examples include having huge areas of highly saturated color, hoping that color coding will be distinguishable in tiny regions, using more nominal categories than the roughly one dozen that can be distinguishable with color coding, or using a sequential scheme for diverging data. Using a red/green hue coding is justifiable only when strong domain conventions exist, and should usually be redundantly coded with luminance differences to be distinguishable to the 10% of men who are color-blind. You should not attempt to visually encode three variables through the three channels of red, green, and blue; they are not separable because they are integrated by the visual system into a combined percept of color. These principles have been clearly explained by many authors, including Ware [41, Chapter 4].

Rainbows Just like in the Sky: The unjustified use of a continuous rainbow colormap is a color pitfall so common that I give it a separate title. The most critical problem is that the standard rainbow colormap is perceptually nonlinear. A fixed range of values that are indistinguishable in the green region would clearly show change in other regions such as where orange changes to yellow or cyan changes to blue. Moreover, hue does not have an implicit perceptual ordering, in contrast to other visual attributes such as greyscale or saturation. If the important aspect of the information to be encoded is low-frequency change, then use a colormap that changes from just one hue to another, or has a single hue that changes saturation. If you are showing high-frequency information, where it is important to distinguish and discuss several nameable regions, then a good strategy is to explicitly quantize your data into a segmented rainbow colormap. These ideas are discussed articulately by Rogowitz and Treinish [32].

4   Late Pitfalls: Paper Strategy, Tactics, and Results

The time to consider the late pitfalls is after the bulk of the project work has been done, but before starting to write your paper draft.

4.1   Strategy Pitfalls

Strategy pitfalls pertain to the content and structure of the entire paper, as opposed to more localized tactics problems that only affect one or a few sections.

What I Did over My Summer Vacation: Do not simply enumerate all activities that required effort when writing a paper. Instead, make judgements about what to discuss in a paper based on your research contributions. This category evokes the grade-school essays many of us were asked to write each fall, which were typically narrative and chronological. Often these Summer Vacation papers contain too much low-level detail, in the extreme cases reading more like a manual than a research paper. For instance, a feature that took weeks or even months of implementation effort may only merit a few sentences, and some features may not be sufficiently interesting to the research community to mention at all. These papers are usually the result of authors who do not know the literature well enough to understand what is and is not a contribution. The solution is to plunge into a more extensive literature review.

Least Publishable Unit: Do not try to squeeze too many papers out of the same project, where you parcel out some tiny increment of research contribution beyond your own previous work. The determination of what is a paper-sized unit of work is admittedly a very individual judgement call, and I will not attempt to define the scope here. As a reviewer, I apply the "I know it when I see it" standard.

Dense as Plutonium: Do not try to cram many papers' worth of content and contributions into one — the inverse pitfall to the one above. These papers are difficult to read because the barrage of ideas is so dense. More importantly, because you don't have enough room to explain the full story of how you have accomplished your work, these papers fail the reproducability test. If you are leaving out so many critical details that the work cannot be replicated by a sufficiently advanced graduate student, then split your writeup into multiple papers.

Bad Slice and Dice: If you have done two papers' worth of work and choose to write two papers, you can still make the wrong choice about how to split up the work between them. In this pitfall, neither paper is truly standalone, yet both repeat too much content found in the other. Repartitioning can solve this problem.

4.2   Tactical Pitfalls

The tactical pitfalls are localized to one or a few sections, as opposed to the paper-level strategy problems above.

Stealth Contributions: Do not leave your contributions implicit or unsaid, whether from intellectual sloth, misplaced modesty, or the hope that the reader may invent a better answer than what you can provide. It is a central piece of your job as an author to clearly and explicitly tell the reader the contributions of your work.

I highly recommend having a sentence near the end of the introduction that starts, "The contribution of this work is", and of using bulleted lists if there are multiple contributions. More subtle ways of stating contributions, using verbs like 'present' and 'propose', can make it more difficult for readers and reviewers to ferret out which of your many sentences is that all-important contributions statement. Also, do not assume that the reader can glean your overall contributions from a close reading of the arguments in your previous work section. While it is critical to have a clear previous work section that states how you address the limitations of the previous work, as I discuss below, your paper should clearly communicate your contributions even if the reader has skipped the entire previous work section.

I find that articulating the contributions requires very careful consideration and is one of the hardest parts of writing up a paper. They are often quite different than the original goals of the project, and often can only be determined in retrospect. What can we do that wasn't possible before? How can we do something better than before? What do we know that was unknown or unclear before? The answers to these questions should guide all aspects of the paper, from the high-level message to the choice of which details are worth discussing. And yet, as an author I find that it's hard to pin these down at the beginning of the writing process. This reason is one of the many to start writing early enough that there is time to refine through multiple drafts. After writing a complete draft, then reading through it critically, I can better refine the contributions spin in a next pass.

I Am So Unique: Do not ignore previous work when writing up your paper. You have to convince the reader that you have done something new, and the only way to do that is to explain how it is different than what has already been done. All research takes place in some kind of intellectual context, and your job as author is to situate what you have done within a framework of that context. A good previous work section is a mini-taxonomy of its own, where you decide on meaningful categorization given your specific topic.

Proposing new names for old techniques or ideas may sneak your work past some reviewers, but will infuriate those who know of that previous work. This tactic will also make you lose credibility with knowledgeable readers. If you cannot find anything related to what you have done, it's more likely that you're looking in the wrong subfield than that your work is a breakthrough of such magnitude that there is no context. Remember to discuss not only work that has been done on similar problems to your own, but also work that uses similar solutions to yours that occurs in different problem domains. This advice is even more critical if you were lax about doing a literature review before you started your project. If you find work similar to your own, you have a fighting chance of carefully differentiating yours in the writeup, but if a reviewer is the one to bring it to your attention, the paper is most likely dead.

Enumeration without Justification: Simply citing the previous work is necessary but not sufficient. A description that "X did Y", even if it includes detail, is not enough. You must explain why this previous work does not itself solve your problem, and what specific limitations of that previous work your approach does address. Every paper you cite in the previous work section is a fundamental challenge to the very existence of your project. Your job is to convince a skeptical reader that the world needs your new thing because it is somehow better than a particular old thing. Moreover, it's not even enough to just make the case that yours is different – yours must be *better*. The claims you make must, of course, be backed up by your validation in a subsequent results section.

A good way to approach the previous work section is that you want to tell to a story to the reader. Figure out the messages you want to get across to the reader, in what order, and then use the references to help you tell this story. It is possible to group the previous work into categories, and to usefully discuss the limitations of the entire category.

Sweeping Assertions: A research paper should not contain sweeping unattributed assertions. You have three choices: cite your source; delete the assertion from your paper; or explicitly tag the statement as your observation, your conjecture, or an explanation of your results. In the last case, the assertion is clearly marked as being part of your research contribution. Be careful with folk wisdom that "everybody knows". You could be mistaken, and tracking down the original sources may change or refine your views. If you cannot find a suitable source after extensive digging, you have stumbled upon a great topic for a future paper! You may either validate and extend the conventional wisdom, or show that it is incorrect.

I Am Utterly Perfect: No work is perfect. An explicit discussion of the limitations of your work strengthens, rather than weakens, your paper. Papers without a discussion of limitations, weaknesses, and implications feel unfinished or preliminary. For instance, how large of a dataset can your system handle? Can you categorize the kinds of datasets for which your technique is suitable and those for which it is not?

4.3   Results Pitfalls

Several pitfalls on how to validate your claims can occur in the results section of your paper.

Unfettered by Time: Do not omit time performance from your writeup, because it is almost always interesting and worth documenting. The level of detail at which you should report this result depends on the paper type and the contribution claims. For instance, a very high-level statement like "interactive response for all datasets shown on a desktop PC" may suffice for an evaluation paper or a design study paper. However, for a technique paper with a contribution claim of better performance than previous techniques, detailed comparison timings in tables or charts would be a better choice.

Fear and Loathing of Complexity: Although most infovis papers do not have detailed proofs of complexity, technique papers that focus on accelerating performance should usually include some statement of algorithm complexity.

Straw Man Comparison: When comparing your technique to previous work, compare against state-of-the-art approaches rather than outdated work. For example, authors unaware of recent work in multilevel approaches to force-directed graph drawing [10] sometimes compare against very naive implementations of spring systems. At the lower level, if you compare benchmarks of your implementation to performance figures quoted from a previous publication and your hardware configuration is more powerful, you should explicitly discuss the difference in capabilities. Better yet, rerun the benchmarks for the competing algorithms on the same machine you use to test your own.

Tiny Toy Datasets: Avoid using only tiny toy datasets in technique papers that refine previously proposed visual encodings. While small synthetic benchmarks can be useful for expository purposes, your validation should include datasets of the same size used by state-of-the-art approaches. Similarly, you should use datasets characteristic of those for your target application.

On the other hand, relatively small datasets may well be appropriate for a user study, if they are carefully chosen in conjunction with some specific target task and this choice is explained and justified.

But My Friends Liked It: Positive informal evaluation of a new infovis system by a few of your infovis-expert labmates is not very compelling evidence that a new technique is useful for novices or scientists in other domains. While the guerilla/discount methodology is great for finding usability problems with products [27], a stronger approach would be informal evaluation with more representative subjects, or formal evaluation with rigorous methodology.

Unjustified Tasks: Beware of running a user study where the tasks are not justified. A study is not very interesting if it shows a nice result for a task that nobody will ever actually do, or a task much less common or important than some other task. You need to convince the reader that your tasks are a reasonable abstraction of the real-world tasks done by your target users. If you are the designer of one of the systems studied, be particularly careful to make a convincing case that you did not cherry-pick tasks with a bias to the strengths of your own system.

5   Final Pitfalls: Style and Submission

After you have a full paper draft, you should check for the final-stage pitfalls.

5.1   Writing Style Pitfalls

Several lower-level pitfalls pertain to writing style.

Deadly Detail Dump: When writing a paper, do not simply dump out all the details and declare victory. The details are the *how* of what you did and do belong at the heart of your paper. But you must first say *what* you did and *why* you did it before the *how*. This advice holds at multiple levels. For the high-level paper structure, start with motivation: why should I, the reader, care about what you've done? Then provide an overview: a big-picture view of what you did. The algorithmic details can then appear after the stage has been set. At the section, subsection, and sometimes even paragraph level, stating the *what* before the *how* will make your writing more clear.

Story-Free Captions: Avoid using a single brusque sentence fragment as your caption text. Caption words are not a precious resource that you should hoard and spend begrudgingly. Instead, design your paper so that as much of the paper story as possible is understandable to somebody who flips through looking only at the figures and captions. Many readers of visualization and graphics papers do exactly this when skimming, so make your captions as standalone as possible.

My Picture Speaks for Itself: You should talk the reader through how your visual representation exposes meaningful structure in the dataset, rather than simply assuming the superiority of your method is obvious to all readers from unassisted inspection of your result images. Technique and design study papers usually include images in the results section showing the visual encodings created by a technique or system on example datasets. The best way to carry out this qualitative evaluation is to compare your method side-by-side with representations created by competing methods on the same dataset.

Grammar Is Optional: Grammar is not optional; you should use correct syntax and punctuation for a smooth low-level flow of words. If English is not your first language, consider having a native speaker check the writing before submitting a paper for review, and also before the final version of your paper goes to press. I recommend Dupré's book [4] as an excellent pitfall-oriented technical writing guide for computer scientists.

Mistakes Were Made: Avoid the passive voice as much as possible. I call out this particular grammar issue because it directly pertains to making your research contribution clear. Is the thing under discussion part of your research contribution, or something that was done or suggested by others? The problem with the passive voice is its ambiguity: the reader does not have enough information to determine *who* did something. This very ambiguity can be the lure of the passive voice to a slothful or overly modest writer. I urge you to use the active voice and make such distinctions explicitly.

Jargon Attack: Avoid jargon as much as possible, and if you must use it then define it first. Definitions are critical both for unfamiliar terms or acronyms, as well as for standard English words being used in a specific technical sense.

Nonspecific Use of Large: Never just use the words 'large' or 'huge' to describe a dataset or the scalability of a technique without giving numbers to clarify the order of magnitude under discussion: hundreds, tens of thousands, millions? Every author has a different idea of what these words mean, ranging from 128 to billions, so be specific. Also, you should provide the size of all datasets used in results figures, so that readers don't have to count dots in an image to guess the numbers.

5.2   Submission Pitfalls

Finally, I caution against pitfalls at the very end of the project, when submitting your paper.

Slimy Simultaneous Submission: Simultaneous submission of the same work at multiple venues who clearly request original work is highly unethical. Moreover, simultaneous submission is stupid, because it is often detected when the same reviewer is independently selected by different conference chairs. The number of experts in any particular subfield can be quite a small set. The standard penalty upon detection is instant dual rejection, and multi-conference blacklists are beginning to be compiled. Finally, even if you do succeed in getting the same work published twice, any gains you make by having a higher publication count will be offset when you lose credibility within your field from those who actually read the work and are annoyed to wade through multiple papers that say the same thing.

Resubmit Unchanged: If your paper is rejected, don't completely ignore the reviews and resubmit to another venue without making any changes. As above, there's a reasonable chance that you'll get the one of the same reviewers again. That reviewer will be highly irritated.

6   Pitfalls By Generality

A cross-cutting way to categorize these pitfalls is by generality. Many hold true for any scientific research paper, rather than being specific to visualization. Of the latter, many hold true for both scientific visualization (scivis) and information visualization (infovis). As many have lamented, the names of these subfields are unfortunate and confusing for outsiders. The definition I use is that it's infovis when the spatial representation is chosen, and it's scivis when the spatial representation is given. Operationally, InfoVis split off as a sister conference from IEEE Visualization (Vis) in 1995. At Vis, the focus is now on scivis.

The choice of paper types is specific to the InfoVis author guide, because this categorization is not explicitly discussed in the Vis call for papers. The first-stage type pitfalls are thus quite specific to infovis. The middle pitfalls on visual encoding are specific to visualization. *Color Cacophony* and *Rainbows Just Like In The Sky* certainly pertain to both infovis and scivis. *Unjustified*

Visual Encoding, Hammer In Search Of Nail, and *2D Good, 3D Better* focus on issues that are more central for an infovis audience, but may well be of benefit to scivis as well. All of the strategy pitfalls pertain to any research paper. The result pitfalls hold for all visualization papers, and *Straw Man Comparison* is general enough for all research papers. The tactical and final stage pitfalls are very general, with two exceptions. *Story-Free Captions* is specific to both visualization and computer graphics. *My Picture Speaks For Itself* is again most tuned for infovis, but certainly may pique the interest of the scivis community.

Although I have framed my discussion in terms of the InfoVis author guide paper categories, my comments also apply to infovis papers in other venues. I argue that even if a call for papers does not explicitly state paper categories, keeping this paper taxonomy in mind will help you write a stronger paper.

7   Conclusion

I have advocated an approach to conducting infovis research that begins with an explicit consideration of paper types. I have exhorted authors to avoid pitfalls at several stages of research process, including visual encoding during design, a checkpoint before starting to write, and after a full paper draft exists. My description and categorization of these pitfalls reflects my own experiences as author, reviewer, and papers chair. I offer it in hopes of steering and stimulating discussion in our field.

Acknowledgments. This paper has grown in part from a series of talks. The impetus to begin articulating my thoughts was the Publishing Your Visualization Research panel at the IEEE Visualization 2006 Doctoral Colloquium. I benefited from discussion with many participants at the 2007 Dagstuhl Seminar on Information Visualization and the 2007 Visual Interactive Effective Worlds workshop at the Lorentz Center, after speaking on the same topic. I also appreciate a discussion with Pat Hanrahan and Lyn Bartram on design studies. John Stasko's thoughts considerably strengthened my discussion of model papers. I thank Aaron Barsky, Hamish Carr, Jeff Heer, Stephen Ingram, Ciarán Llachlan Leavitt, Peter McLachlan, James Slack, and Matt Ward for feedback on paper drafts. I particularly thank Heidi Lam, Torsten Möller, John Stasko, and Melanie Tory for extensive discussions beyond the call of duty.

References

1. Amar, R., Eagan, J., Stasko, J.: Low-level components of analytic activity in information visualization. In: Proc. IEEE Symposium on Information Visualization (InfoVis), pp. 111–117 (2005)
2. Becker, R.A., Cleveland, W.S., Shyu, M.J.: The visual design and control of trellis display. Journal of Computational and Statistical Graphics 5, 123–155 (1996)
3. Card, S., Mackinlay, J.: The structure of the information visualization design space. In: Proc. IEEE Symposium on Information Visualization (InfoVis), pp. 92–99 (1997)

4. Dupré, L.: Bugs in Writing. Addison-Wesley, Reading (1995)
5. Fekete, J.-D., Plaisant, C.: Interactive information visualization of a million items. In: Proc. IEEE Symposium on Information Visualization (InfoVis), pp. 117–126 (2002)
6. Fua, Y.-H., Ward, M.O., Rundensteiner, E.A.: Hierarchical parallel coordinates for visualizing large multivariate data sets. In: Proc. IEEE Visualization Conf. (Vis), pp. 43–50 (1999)
7. Furnas, G., Bederson, B.: Space-scale diagrams: Understanding multiscale interfaces. In: Proc. ACM Conf. Human Factors in Computing Systems (CHI), pp. 234–241 (1995)
8. Furnas, G.W.: A fisheye follow-up: Further reflection on focus + context. In: Proc. ACM Conf. Human Factors in Computing Systems (CHI), pp. 999–1008 (2006)
9. Grossman, T., Wigdor, D., Balakrishnan, R.: Exploring and reducing the effects of orientation on text readability in volumetric displays. In: Proc. ACM Conf. Human Factors in Computing Systems (CHI), pp. 483–492 (2007)
10. Hachul, S., Jünger, M.: Drawing large graphs with a potential-field-based multilevel algorithm. In: Pach, J. (ed.) GD 2004. LNCS, vol. 3383, pp. 285–295. Springer, Heidelberg (2005)
11. Havre, S., Hetzler, B., Nowell, L.: Themeriver(tm): In search of trends, patterns, and relationships. In: Proc. IEEE Symposium on Information Visualization (InfoVis), pp. 115–123 (2000)
12. Healey, C.G.: Perception in visualization, cited 14 Nov. 2007, http://www.csc.ncsu.edu/faculty/healey/PP
13. Heer, J., boyd, d.: Vizster: Visualizing online social networks. In: Proc. IEEE Symposium on Information Visualization (InfoVis), pp. 32–39 (2005)
14. Heer, J., Card, S.K., Landay, J.A.: prefuse: a toolkit for interactive information visualization. In: Proc. ACM Conf. Human Factors in Computing Systems (CHI), pp. 421–430 (2005)
15. Holten, D.: Hierarchical edge bundles: Visualization of adjacency relations in hierarchical data. IEEE Trans. Visualization and Computer Graphics (TVCG) (Proc. InfoVis 06) 12(5), 741–748 (2006)
16. Hornbæk, K., Hertzum, M.: Untangling the usability of fisheye menus. ACM Trans. on Computer-Human Interaction (ToCHI) 14(2), article 6 (2007)
17. Isaacs, E., Tang, J.: Why don't more non-North-American papers get accepted to CHI? SIGCHI Bulletin 28(1) (1996), http://www.sigchi.org/bulletin/1996.1/isaacs.html
18. Johnson, C., Moorhead, R., Munzner, T., Pfister, H., Rheingans, P., Yoo, T.S.: NIH/NSF Visualization Research Challenges Report. IEEE Computer Society Press, Los Alamitos (2006)
19. Johnson, R.E., et al.: How to get a paper accepted at OOPSLA. In: Proc. Conf. Object Oriented Programming Systems Languages and Applications (OOPSLA), pp. 429–436 (1996), http://www.sigplan.org/oopsla/oopsla96/how93.html
20. Kajiya, J.: How to get your SIGGRAPH paper rejected, http://www.siggraph.org/publications/instructions/rejected
21. Kincaid, R., Ben-Dor, A., Yakhini, Z.: Exploratory visualization of array-based comparative genomic hybridization. Information Visualization 4(3), 176–190 (2005)

22. Levin, R., Redell, D.D.: An evaluation of the ninth SOSP submissions; or, how (and how not) to write a good systems paper. ACM SIGOPS Operating Systems Review 17(3), 35–40 (1983),
    `http://www.usenix.org/events/samples/submit/advice.html`
23. MacEachren, A., Dai, X., Hardisty, F., Guo, D., Lengerich, G.: Exploring high-D spaces with multiform matrices and small multiples. In: Proc. IEEE Symposium on Information Visualization (InfoVis), pp. 31–38 (2003)
24. Mackinlay, J.D.: Automating the Design of Graphical Presentations of Relational Information. ACM Trans. on Graphics (TOG) 5(2), 111–141 (1986)
25. Munzner, T.: Drawing large graphs with H3Viewer and Site Manager. In: Whitesides, S.H. (ed.) GD 1998. LNCS, vol. 1547, pp. 384–393. Springer, Heidelberg (1999)
26. Munzner, T., Burchard, P.: Visualizing the structure of the world wide web in 3D hyperbolic space. In: Proc. Virtual Reality Modelling Language Symposium (VRML), pp. 33–38. ACM SIGGRAPH (1995)
27. Nielsen, J.: Guerrilla HCI: Using discount usability engineering to penetrate the intimidation barrier. In: Bias, R.G., Mayhew, D.J. (eds.) Cost-justifying usability, pp. 245–272. Academic Press, London (1994)
28. Partridge, C.: How to increase the chances your paper is accepted at ACM SIGCOMM. ACM SIGCOMM Computer Communication Review 28(3) (1998),
    `http://www.sigcomm.org/conference-misc/author-guide.html`
29. Plaisant, C.: The challenge of information visualization evaluation. In: Proc. Advanced Visual Interfaces (AVI), pp. 109–116 (2004)
30. Plumlee, M., Ware, C.: Zooming versus multiple window interfaces: Cognitive costs of visual comparisons. Proc. ACM Trans. on Computer-Human Interaction (ToCHI) 13(2), 179–209 (2006)
31. Pousman, Z., Stasko, J.T., Mateas, M.: Casual information visualization: Depictions of data in everyday life. IEEE Trans. Visualization and Computer Graphics (TVCG) (Proc. InfoVis 07) 13(6), 1145–1152 (2007)
32. Rogowitz, B.E., Treinish, L.A.: How not to lie with visualization. Computers In Physics 10(3), 268–273 (1996),
    `http://www.research.ibm.com/dx/proceedings/pravda/truevis.htm`
33. Shaw, M.: Mini-tutorial: Writing good software engineering research papers. In: Proc. Intl. Conf. on Software Engineering (ICSE), pp. 726–736 (2003),
    `http://www.cs.cmu.edu/~Compose/shaw-icse03.pdf`
34. Shewchuk, J.: Three sins of authors in computer science and math (1997),
    `http://www.cs.cmu.edu/jrs/sins.html`
35. Shneiderman, B., Plaisant, C.: Strategies for evaluating information visualization tools: Multi-dimensional in-depth long-term case studies. In: Proc. AVI Workshop on BEyond time and errors: novel evaLuation methods for Information Visualization (BELIV), pp. 38–43 (2006)
36. Smallman, H.S., John, M.S., Oonk, H.M., Cowen, M.B.: Information availability in 2D and 3D displays. IEEE Computer Graphics and Applications (CG&A) 21(5), 51–57 (2001)
37. Tang, D., Stolte, C., Bosch, R.: Design choices when architecting visualizations. Information Visualization 3(2), 65–79 (2004)
38. Tory, M., Kirkpatrick, A.E., Atkins, M.S., Möller, T.: Visualization task performance with 2D, 3D, and combination displays. IEEE Trans. Visualization and Computer Graphics (TVCG) 12(1), 2–13 (2006)

39. Trafton, J.G., Kirschenbaum, S.S., Tsui, T.L., Miyamoto, R.T., Ballas, J.A., Raymond, P.D.: Turning pictures into numbers: Extracting and generating information from complex visualizations. Intl. Journ. Human Computer Studies 53(5), 827–850 (2000)
40. van Wijk, J.J., van Selow, E.R.: Cluster and calendar based visualization of time series data. In: Proc. IEEE Symposium on Information Visualization (InfoVis), pp. 4–9 (1999)
41. Ware, C.: Information Visualization: Perception for Design, 2nd edn. Morgan Kaufmann/Academic Press, London (2004)
42. Weaver, C., Fyfe, D., Robinson, A., Holdsworth, D.W., Peuquet, D.J., MacEachren, A.M.: Visual analysis of historic hotel visitation patterns. Information Visualization 6(1), 89–103 (2007)
43. Wilkinson, L., Anand, A., Grossman, R.: Graph-theoretic scagnostics. In: Proc. IEEE Symposium on Information Visualization (InfoVis), pp. 157–164 (2005)
44. Yost, B., North, C.: The perceptual scalability of visualization. IEEE Trans. Visualization and Computer Graphics (TVCG) (Proc. InfoVis 06) 12(5), 837–844 (2006)

Visual Analytics:
Definition, Process, and Challenges

Daniel Keim[1], Gennady Andrienko[2], Jean-Daniel Fekete[3], Carsten Görg[4],
Jörn Kohlhammer[5], and Guy Melançon[6]

[1] Department of Computer and Information Science, University of Konstanz,
78457 Konstanz, Germany,
keim@informatik.uni-konstanz.de
[2] Fraunhofer Institute for Intelligent Analysis and Information Systems(IAIS),
Schloss Birlinghoven 53754 Sankt Augustin, Germany,
gennady.andrienko@iais.fraunhofer.de
[3] Université Paris-Sud, INRIA, Bât 490,
F-91405 Orsay Cedex, France,
Jean-Daniel.Fekete@inria.fr
[4] School of Interactive Computing & GVU Center, Georgia Institute of Technology,
85 5th St., NW, Atlanta, GA 30332-0760, USA,
goerg@cc.gatech.edu
[5] Fraunhofer Institute for Computer Graphics Research,
Fraunhoferstraße 5, D-64283 Darmstadt, Germany,
joern.kohlhammer@igd.fraunhofer.de
[6] INRIA Bordeaux – Sud-Ouest, CNRS UMR 5800 LaBRI,
Campus Université Bordeaux I,
351 Cours de la libération, 33405 Talence Cedex, France,
Guy.Melancon@labri.fr

1 Introduction and Motivation

We are living in a world which faces a rapidly increasing amount of data to be dealt with on a daily basis. In the last decade, the steady improvement of data storage devices and means to create and collect data along the way influenced our way of dealing with information: Most of the time, data is stored without filtering and refinement for later use. Virtually every branch of industry or business, and any political or personal activity nowadays generate vast amounts of data. Making matters worse, the possibilities to collect and store data increase at a faster rate than our ability to use it for making decisions. However, in most applications, raw data has no value in itself; instead we want to extract the information contained in it.

The **information overload problem** refers to the danger of getting lost in data which may be

- irrelevant to the current task at hand
- processed in an inappropriate way
- presented in an inappropriate way

A. Kerren et al. (Eds.): Information Visualization, LNCS 4950, pp. 154–175, 2008.

Due to information overload, time and money are wasted, scientific and industrial opportunities are lost because we still lack the ability to deal with the enormous data volumes properly. People in both their business and private lives, decision-makers, analysts, engineers, emergency response teams alike, are often confronted with massive amounts of disparate, conflicting and dynamic information, which are available from multiple heterogeneous sources. We want to simply and effectively exploit and use the hidden opportunities and knowledge resting in unexplored data sources.

In many application areas success depends on the right information being available at the right time. Nowadays, the acquisition of raw data is no longer the driving problem: It is the ability to identify methods and models, which can turn the data into reliable and provable knowledge. Any technology, that claims to overcome the information overload problem, has to provide answers for the following problems:

– Who or what defines the "relevance of information" for a given task?
– How can appropriate procedures in a complex decision making process be identified?
– How can the resulting information be presented in a decision- or task-oriented way?
– What kinds of interaction can facilitate problem solving and decision making?

With every new "real-life" application, procedures are put to the test possibly under circumstances completely different from the ones under which they have been established. The awareness of the problem how to understand and analyse our data has been greatly increased in the last decade. Even as we implement more powerful tools for automated data analysis, we still face the problem of understanding and "analysing our analyses" in the future: Fully-automated search, filter and analysis only work reliably for well-defined and well-understood problems. The path from data to decision is typically quite complex. Even as fully-automated data processing methods represent the knowledge of their creators, they lack the ability to communicate their knowledge. This ability is crucial: If decisions that emerge from the results of these methods turn out to be wrong, it is especially important to examine the procedures.

The overarching driving vision of **visual analytics** is to turn the information overload into an opportunity: Just as *information visualization* has changed our view on databases, the goal of Visual Analytics is to make *our way of processing* data and information transparent for an analytic discourse. The visualization of these processes will provide the means of communicating about them, instead of being left with the results. Visual Analytics will foster the constructive evaluation, correction and rapid improvement of our processes and models and - ultimately - the improvement of our knowledge and our decisions (see Figure 1).

On a grand scale, visual analytics solutions provide technology that combines the strengths of human and electronic data processing. Visualization becomes the medium of a semi-automated analytical process, where humans and machines cooperate using their respective distinct capabilities for the most effective results.

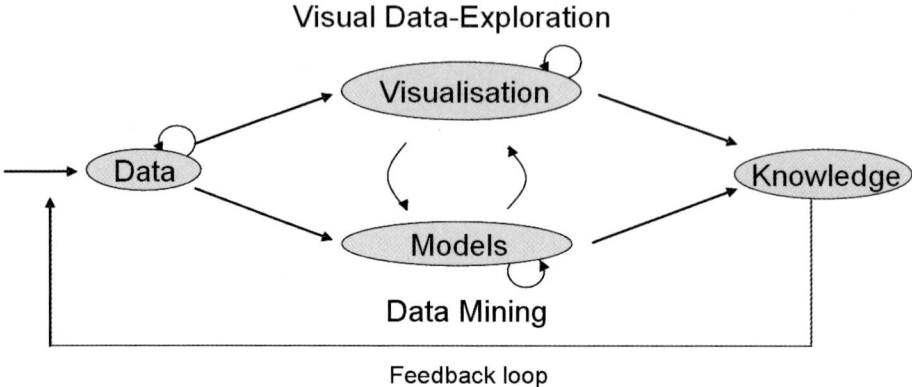

Fig. 1. Tight integration of visual and automatic data analysis methods with database technology for a scalable interactive decision support.

The user has to be the ultimate authority in giving the direction of the analysis along his or her specific task. At the same time, the system has to provide effective means of interaction to concentrate on this specific task. On top of that, in many applications different people work along the path from data to decision. A visual representation will sketch this path and provide a reference for their collaboration across different tasks and abstraction levels.

The diversity of these tasks can not be tackled with a single theory. Visual analytics research is highly interdisciplinary and combines various related research areas such as visualization, data mining, data management, data fusion, statistics and cognition science (among others). Visualization has to continuously challenge the perception by many of the applying sciences that visualization is not a scientific discipline in its own right. Even if the awareness exists, that scientific analysis and results must be visualized in one way or the other, this often results in ad hoc solutions by application scientists, which rarely match the state of the art in interactive visualization science, much less the full complexity of the problems. In fact, all related research areas in the context of visual analytics research conduct rigorous, serious science each in a vibrant research community. To increase the awareness of their work and their implications for visual analytics research clearly emerges as one main goal of the international visual analytics community (see Figure 2).

Because visual analytics research can be regarded as an integrating discipline, application specific research areas should contribute with their existing procedures and models. Emerging from highly application-oriented research, dispersed research communities worked on specific solutions using the repertoire and standards of their specific fields. The requirements of visual analytics introduce new dependencies between these fields.

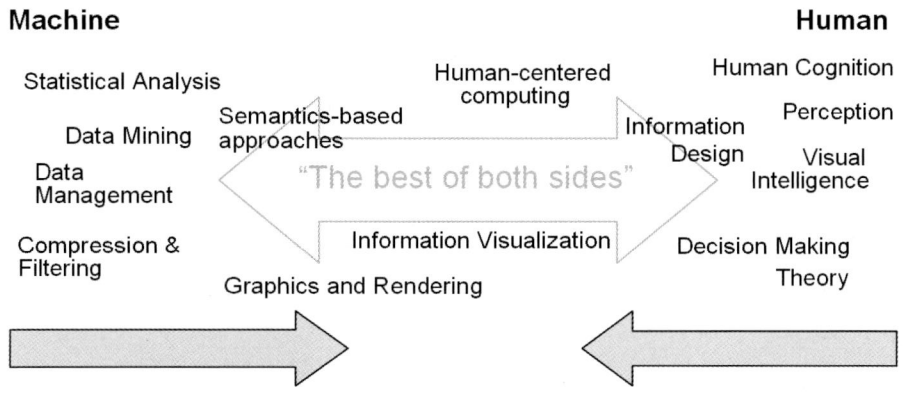

Fig. 2. Visual analytics integrates scientific disciplines to improve the division of labor between human and machine.

2   Definition of Visual Analytics

In "Illuminating the Path" [39], Thomas and Cook define visual analytics as the science of analytical reasoning facilitated by interactive visual interfaces. In this paper, however, we would like to give a more specific definition:

Visual analytics combines automated analysis techniques with interactive visualizations for an effective understanding, reasoning and decision making on the basis of very large and complex data sets.

The goal of visual analytics is the creation of tools and techniques to enable people to:

- Synthesize information and derive insight from massive, dynamic, ambiguous, and often conflicting data.
- Detect the expected and discover the unexpected.
- Provide timely, defensible, and understandable assessments.
- Communicate assessment effectively for action.

By integrating selected science and technology from the above discussed disciplines and as illustrated in Figure 2, there is the promising opportunity to form the unique and productive field of visual analytics. Work in each of the participating areas focuses on different theoretical and practical aspects of users solving real-world problems using Information Technology in an effective and efficient way. These areas have in common similar scientific challenges and significant scientific added-value from establishing close collaboration can be identified. Benefit of collaboration between the fields is identified to be two-fold:

- Jointly tackling common problems will arrive at better results on the local level of each discipline, in a more efficient way.
- Integrating appropriate results from each of the disciplines will lay the fundament for significantly improved solutions in many important data analysis applications.

Visual Analytics versus Information Visualization

Many people are confused by the new term visual analytics and do not see a difference between the two areas. While there is certainly some overlay and some of the information visualization work is certainly highly related to visual analytics, traditional visualization work does not necessarily deal with an analysis tasks nor does it always also use advanced data analysis algorithms.

Visual analytics is more than just visualization. It can rather be seen as an integral approach to decision-making, combining visualization, human factors and data analysis. The challenge is to identify the best automated algorithm for the analysis task at hand, identify its limits which can not be further automated, and then develop a tightly integrated solution with adequately integrates the best automated analysis algorithms with appropriate visualization and interaction techniques.

While some of such research has been done within the visualization community in the past, the degree to which advanced knowledge discovery algorithms have been employed is quite limited. The idea of visual analytics is to fundamentally change that. This will help to focus on the right part of the problem, i.e. the parts that can not be solved automatically, and will provide solutions to problems that we were not able to solve before.

One important remark should be made here. Most research efforts in Information Visualization have concentrated on the process of producing views and creating valuable interaction techniques for a given class of data (social network, multi-dimensional data, etc.). However, much less has been suggested as to how user interactions on the data can be turned into intelligence to tune underlying analytical processes. A system might for instance observe that most of the user's attention concern only a subpart of an ontology (through queries or by repeated direct manipulations of the same graphical elements, for instance). Why not then use this knowledge about the user's interest and update various parameters by the system (trying to systematically place elements or components of interest in center view, even taking this fact into account when driving a clustering algorithm with a modularity quality criteria, for instance).

This is one place where Visual Analytics maybe differs most from Information Visualization, giving higher priority to data analytics from the start and through all iterations of the sense making loop. Creativity is then needed to understand how perception issues can help bring more intelligence into the analytical process by "learning" from users' behavior and effective use of the visualization.

3   Areas Related to Visual Analytics

Visual analytics builds on a variety of related scientific fields. At its heart, Visual Analytics integrates Information and Scientific Visualization with Data Management and Data Analysis Technology, as well as Human Perception and Cognition research. For effective research, Visual Analytics also requires an appropriate Infrastructure in terms of software and data sets and related analytical problems repositories, and to develop reliable Evaluation methodology (see Figure 3).

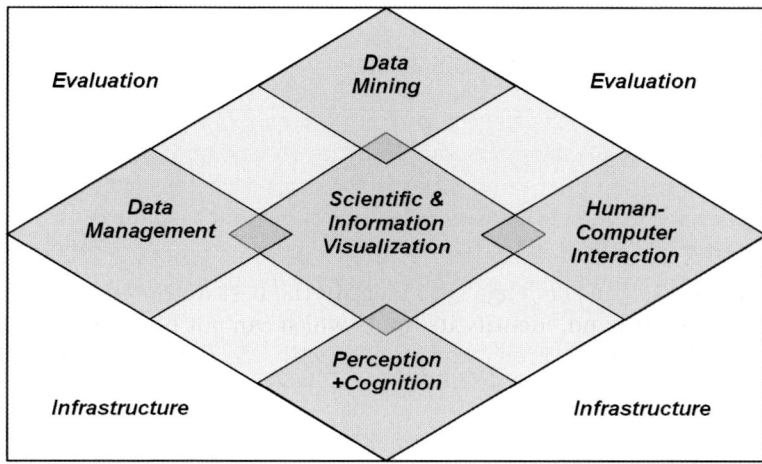

Fig. 3. Visual Analytics integrates Scientific and Information Visualization with core adjacent disciplines: Data management and analysis, spatio-temporal data, and human perception and cognition. Successful Visual Analytics research also depends on the availability of appropriate infrastructure and evaluation facilities.

An example for a common problem in several of the disciplines is that of scalability with data size. The larger the data set to be handled gets, the more difficult it gets to manage, analyze, and visualize these data effectively. Researching appropriate forms to represent large data volumes by smaller volumes containing the most relevant information benefits each of the data management, analysis, and visualization fields. On top of these individual progresses, a synergetic collaboration of all these fields may lead to significantly improved processing results. Consider a very large data stream. Appropriate data management technology gives efficient access to the stream, which is intelligently processed and abstracted by an automatic analysis algorithm which has an interface to the data management layer. On top of the analysis output, an interactive visualization which is optimized for efficient human perception of the relevant information allows the analyst to consume the analysis results, and adapt relevant parameters of the data aggregation an analysis engines as appropriate. The combination of the individual data handling steps into a Visual Analytics pipeline leads to improved results and makes data domains accessible which are not effectively accessible by any of the individual data handling disciplines. Similar argumentations apply to other related fields and disciplines. In many fields, visualization is already used and developed independently as a means for analyzing the problems at hand. However, a unified, interdisciplinary perspective on using visualization for analytical problem-solving will show beneficial for all involved disciplines. As common principles, best practices, and theories will be developed, these will become usable in the individual disciplines and application domains, providing economies of scale, avoiding replication of work or application of only sub-optimal techniques.

3.1 Visualization

Visualization has emerged as a new research discipline during the last two decades. It can be broadly classified into *Scientific* and *Information* Visualization.

In Scientific Visualization, the data entities to be visualized are typically 3D geometries or can be understood as scalar, vectorial, or tensorial fields with explicit references to time and space. A survey of current visualization techniques can be found in [22,35,23]. Often, 3D scalar fields are visualized by isosurfaces or semi-transparent point clouds (direct volume rendering) [15]. To this end, methods based on optical emission- or absorption models are used which visualize the volume by ray-tracing or projection. Also, in the recent years significant work focused on the visualization of complex 3-dimensional flow data relevant e.g., in aerospace engineering [40]. While current research has focused mainly on efficiency of the visualization techniques to enable interactive exploration, more and more methods to automatically derive relevant visualization parameters come into focus of research. Also, interaction techniques such as focus&context [28] gain importance in scientific visualization.

Information Visualization during the last decade has developed methods for the visualization of abstract data where no explicit spatial references are given [38,8,24,41]. Typical examples include business data, demographics data, network graphs and scientific data from e.g., molecular biology. The data considered often comprises hundreds of dimensions and does not have a natural mapping to display space, and renders standard visualization techniques such as (x, y) plots, line- and bar-charts ineffective. Therefore, novel visualization techniques are being developed by employing e.g., Parallel Coordinates and their numerous extensions [20], Treemaps [36], and Glyph [17]- and Pixel-based [25] visual data representations. Data with inherent network structure may be visualized using graph-based approaches. In many Visualization application areas, the typically huge volumes of data require the appropriate usage of automatic data analysis techniques such as clustering or classification as preprocessing prior to visualization. Research in this direction is just emerging.

3.2 Data Management

An efficient management of data of various types and qualities is a key component of Visual Analytics as this technology typically provides the input of the data which are to be analyzed. Generally, a necessary precondition to perform any kind of data analysis is an integrated and consistent data basis [18,19]. Database research has until the last decade focused mainly on aspects of efficiency and scalability of exact queries on homogeneous, structured data. With the advent of the Internet and the easy access it provides to all kinds of heterogeneous data sources, the database research focus has shifted toward integration of heterogeneous data. Finding integrated representation of different data types such as numeric data, graphs, text, audio and video signals, semi-structured data, semantic representations and so on is a key problem of modern database technology. But the availability of heterogeneous data not only requires the mapping of database schemata but includes also the cleaning and harmonization of uncertainty and missing data in the volumes of heterogeneous data. Modern applications require such intelligent data fusion to be feasible in near real-time and as automatically as possible [32]. New forms of information sources such as data streams [11], sensor networks [30] or automatic extraction of information from large document collections (e.g., text, HTML) result in a difficult data analysis problem which to support is currently in the focus of database research [43]. The relationship between Data Management, Data Analysis and Visualization is characterized such that Data Management techniques developed increasingly rely on intelligent data analysis techniques, and also interaction and visualization to arrive at optimal results. On the other hand, modern database systems provide the input data sources which are to be visually analyzed.

3.3   Data Analysis

Data Analysis (also known as Data Mining or Knowledge Discovery) researches methods to automatically extract valuable information from raw data by means of automatic analysis algorithms [29,16,31]. Approaches developed in this area can be best described by the addressed analysis tasks. A prominent such task is supervised learning from examples: Based on a set of training samples, deterministic or probabilistic algorithms are used to learn models for the classification (or prediction) of previously unseen data samples [13]. A huge number of algorithms have been developed to this end such as Decision Trees, Support Vector Machines, Neuronal Networks, and so on. A second prominent analysis task is that of cluster analysis [18,19], which aims to extract structure from data without prior knowledge being available. Solutions in this class are employed to automatically group data instances into classes based on mutual similarity, and to identify outliers in noisy data during data preprocessing for subsequent analysis steps. Further data analysis tasks include tasks such as association rule mining (analysis of co-occurrence of data items) and dimensionality reduction. While data analysis initially was developed for structured data, recent research aims at analyzing also semi-structured and complex data types such as web documents or multimedia data [34].

It has recently been recognized that visualization and interaction are highly beneficial in arriving at optimal analysis results [9]. In almost all data analysis algorithms a variety of parameters needs to be specified, a problem which is usually not trivial and often needs supervision by a human expert. Visualization is also a suitable means for appropriately communicating the results of the automatic analysis, which often is given in abstract representation, e.g., a decision tree. Visual Data Mining methods [24] try to achieve exactly this.

3.4   Perception and Cognition

Effective utilization of the powerful human perception system for visual analysis tasks requires the careful design of appropriate human-computer interfaces. Psychology, Sociology, Neurosciences and Design each contribute valuable results to the implementation of effective visual information systems. Research in this area focuses on user-centered analysis and modeling (Requirement Engineering), the development of principles, methods and tools for design of perception-driven, multimodal interaction techniques for visualization and exploration of large information spaces, as well as usability evaluation of such systems [21,12]. On the technical side, research in this area is influenced by two main factors: (1.) The availability of improved display resources (hardware), and (2.) Development of novel interaction algorithms incorporating machine recognition of the actual user intent and appropriate adaptation of main display parameters such as the level of detail, data selection and aggregation, etc. by which the data is presented[44]. Important problems addressed in this area include the research of perceptual, cognitive and graphical principles which in combination lead to improved visual communication of data and analysis results; The development of perception-theory-based solutions for the graphical representation of static and dynamic structures; And development of visual representation of information at several levels of abstraction, and optimization of existing focus-and-context techniques.

3.5   Human-Computer Interaction

Human-computer interaction is the research area that studies the interaction between people and computers. It involves the design, implementation and evaluation of interactive systems in the context of the user's task and work [12]. Like visual analytics itself, human-computer interaction is a multi-disciplinary research area that draws on many other disciplines: computer science, system design, and behavioral science are some of them. The basic underlying research goal is to improve the interaction between users and computers: how to make computers more receptive to the users' intentions and needs. Thus, the research areas discussed in the previous section about perception and cognition are also much related to human-computer interaction [21].

As pointed out in the introduction, visual analytics aims to combine and integrate the strengths of computers and humans into an interactive process to extract knowledge from data. To effectively switch back and forth between tasks for the computer and tasks for the human it is crucial to develop an effective user interface that minimizes the barrier between the human's cognitive model of what they want to accomplish and the computer's understanding of the human's task. The design of user interfaces focuses on human factors of interactive software, methods to develop and assess interfaces, interaction styles, and design considerations such as effective messages and appropriate color choice [37].

3.6   Infrastructure and Evaluation

The above described research disciplines require cross-discipline support regarding the evaluation of the found solutions, and need certain infrastructure and standardization grounding to build on effectively. In the field of information visualization, standardization and evaluation came into the focus of research only recently. It has been realized that a general understanding of the taxonomies regarding the main data types and user tasks [2] to be supported are highly desirable for shaping visual analytics research. A common understanding of data and problem dimensions and structure, and acceptance of evaluation standards will make research results better comparable, optimizing research productivity. Also, there is an obvious need to build repositories of available analysis and visualization algorithms, which researchers can build upon in their work, without having to re-implement already proven solutions.

How to assess the value of visualization is a topic of lively debate [42,33]. A common ground that can be used to position and compare future developments in the field of data analysis is needed. The current diversification and dispersion of visual analytics research and development resulted from its focus onto specific application areas. While this approach may suit the requirements of each of these applications, a more rigorous and overall scientific perspective will lead to a better understanding of the field and a more effective and efficient development of innovative methods and techniques.

3.7   Sub-communities

Spatio-Temporal Data: While many different data types exist, one of the most prominent and ubiquitous data types is data with references to time and space. The importance of this data type has been recognized by a research community which formed around spatio-temporal data management and analysis [14]. In geospatial data research, data with references in the real world coming from e.g., geographic measurements, GPS position data, remote sensing applications, and so on is considered. Finding spatial relationships and patterns among this data is of special interest, requiring the development of appropriate management, representation and analysis functions. E.g., developing efficient data structures or defining distance and similarity functions is in the focus of research. Visualization often plays a key role in the successful analysis of geospatial data [6,26].

In temporal data, the data elements can be regarded as a function of time. Important analysis tasks here include the identification of patterns (either linear or periodical), trends and correlations of the data elements over time, and application-dependent analysis functions and similarity metrics have been proposed in fields such as finance, science, engineering, etc. Again, visualization of time-related data is important to arrive at good analysis results [1].

The analysis of data with references both in space and in time is a challenging research topic. Major research challenges include [4]: scale, as it is often necessary to consider spatio-temporal data at different spatio-temporal scales; the uncertainty of the data as data are often incomplete, interpolated, collected at different times, or based upon different assumptions; complexity of geographical space and time, since in addition to metric properties of space and time and topological/temporal relations between objects, it is necessary to take into account the heterogeneity of the space and structure of time; and complexity of spatial decision making processes, because a decision process may involve heterogeneous actors with different roles, interests, levels of knowledge of the problem domain and the territory.

Network and Graph Data: Graphs appear as flexible and powerful mathematical tools to model real-life situations. They naturally map to transportation networks, electric power grids, and they are also used as artifacts to study complex data such as observed interactions between people, or induced interactions between various biological entities. Graphs are successful at turning semantic proximity into topological connectivity, making it possible to address issues based on algorithmics and combinatorial analysis.

Graphs appear as essential modeling and analytical objects, and as effective visual analytics paradigms. Major research challenges are to produce scalable analytical methods to identify key components both structurally and visually. Efforts are needed to design process capable of dealing with large datasets while producing readable and usable graphical representations, allowing proper user interaction. Special efforts are required to deal with dynamically changing networks, in order to assess of structural changes at various scales.

4   The Visual Analytics Process

A number of systems for information visualization, as well as specific visualization techniques, motivate their design choice from Shneiderman's celebrated mantra "Overview first, Filter and zoom, Details on demand". As is, the mantra clearly emphasizes the role of visualization in the knowledge discovery process. Recently, Keim adjusted the mantra to bring its focus toward Visual Analytics: "Analyze first, Show the Important, Zoom, filter and analyze further, Details on demand". In other words, this mantra is calling for astute combinations of analytical approaches together with advanced visualization techniques.

The computation of any visual representation and/or geometrical embedding of large and complex datasets requires some analysis to start with. Many scalable graph drawing algorithms try to take advantage of any knowledge on topology to optimize the drawing in terms of readability. Other approaches offer representations composed of visual abstractions of clusters to improve readability. The challenge then is to try to come up with a representation that is as faithful as possible to avoid introducing uncertainty. We must not fall into the naïve assumption that visualization can offer a virgin view on the data: any representation will inevitably favor an interpretation over all possible ones. The solution offered by Visual Analytics is then to let the user enter into a loop where data can be interactively manipulated to help gain insight both on the data and the representation itself.

The sense-making loop structures the whole knowledge discovery process supported through Visual Analytics. A generic scenario can be given following a schema developed by van Wijk [42], which furthermore admits to be evaluated and measured in terms of efficiency or knowledge gained. A choice for an initial representation and adequate interactions can be made after applying different statistical and mathematical techniques, such as spatio-temporal data analysis or link mining depending on the nature of the dataset under study. The process then enters a loop where the user can gain knowledge on the data, ideally driving the system toward more focused and more adequate analytical techniques. Dually, interacting on the visual representation, the user will gain a better understanding of the visualization itself commanding for different views helping him or her to go beyond the visual and ultimately confirm hypotheses built from previous iterations (see Figure 4).

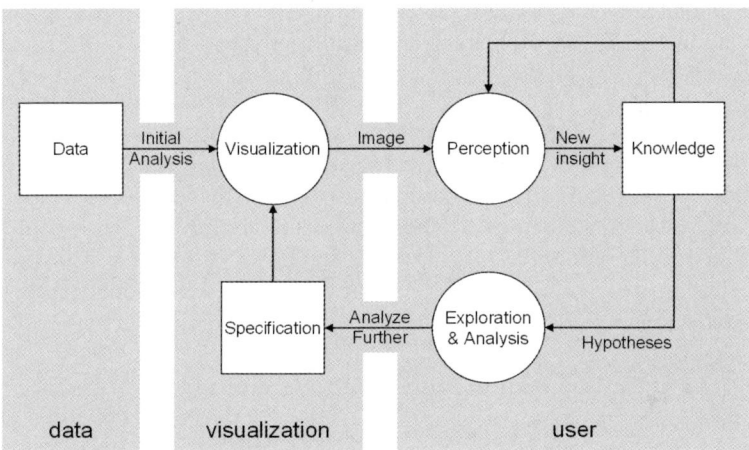

Fig. 4. The sense-making loop for Visual Analytics based on the simple model of visualization by Wijk [42].

5   Application Challenges

Visual Analytics is a highly application oriented discipline driven by practical requirements in important domains. Without attempting a complete survey over all possible application areas, we sketch the potential applicability of Visual Analytics technology in a few key domains.

In the **Engineering** domain, Visual Analytics can contribute to speed-up development time for products, materials, tools and production methods by offering more effective, intelligent access to the wealth of complex information resulting from prototype development, experimental test series, customers' feedback, and many other performance metrics. One key goal of applied Visual Analytics in the engineering domain will be the analysis of the complexity of the production systems in correlation with the achieved output, for an efficient and effective improvement of the production environments.

Financial Analysis is a prototypical promising application area for Visual Analytics. Analysts in this domain are confronted with streams of heterogeneous information from different sources available at high update rates, and of varying reliability. Arriving at a unifying, task-centered view on diverse streams of data is a central goal in financial information systems. Integrated analysis and visualization of heterogeneous data types such as news feeds, real-time trading data, and fundamental economic indicators poses a challenge for developing advanced analysis solutions in this area. Research based on results from Information Visualization is regarded as promising in this case.

Socio-economic considerations often form the basis of political decision processes. A modern society can be regarded as a complex system of interrelationships between political decisions and economic, cultural and demographic effects. Analysis and Visualization of these interrelationships is promising in developing a better understanding of these phenomena, and to arrive at better decisions. Successful Visual Analytics applications in this domain could start being developed based on currently existing Geo-Spatial analysis frameworks.

Public **Safety & Security** is another important application area where Visual Analytics may contribute with advanced solutions. Analysts need to constantly monitor huge amounts of heterogeneous information streams, correlating information of varying degrees of abstraction and reliability, assessing the current level of public safety, triggering alert in case of alarming situations being detected. Data integration and correlation combined with appropriate analysis and interactive visualization is promising to develop more efficient tools for the analysis in this area.

The study of **Environment and Climate change** often requires the examination of long term weather records and logs of various sensors, in a search for patterns that can be related to observations such as changes in animal populations, or in meteorological and climatic processes for instance. These requirements call for the development of systems allowing visual and graphical access to historical monitoring data and predictions from various models in search for or in order to validate patterns building over time.

These diverse fields of applications share many problems on an abstract level, most of which are addressed by Visual Analytics. The actual (software) solution must be adapted to the specific needs and terminologies of the application area and consequently, many researchers currently focus on a specific customer segment. Much can be achieved, if the European research infrastructure in this field becomes strong enough to encourage the exchange of ideas on a broad scale, to foster development of solutions applicable to multiple domains, achieving synergy effects.

6   Technical Challenges

The primary goal of Visual Analytics is the analysis of vast amounts of data to identify and visually distill the most valuable and relevant information content. The visual representation should reveal structural patterns and relevant data properties for easy perception by the analyst. A number of key requirements need to be addressed by advanced Visual Analytics solutions. We next outline important scientific challenges in this context.

Scalability with Data Volumes and Data Dimensionality: Visual Analytics techniques need to be able to scale with the size and dimensionality of the input data space. Techniques need to accommodate and graphically represent high-resolution input data as well as continuous input data streams of high bandwidth. In many applications, data from multiple, heterogeneous sources need to be integrated and processed jointly. In these cases, the methods need to be able to scale with a range of different data types, data sources, and levels of quality. The visual representation algorithms need to be efficient enough for implementation in interactive systems.

Quality of Data and Graphical Representation: A central issue in Visual Analytics is the avoidance of misinterpretations by the analyst. This may result due to uncertainty and errors in the input data, or limitations of the chosen analysis algorithm, and may produce misleading analysis results. To face this problem, the notion of data quality, and the confidence of the analysis algorithm needs to be appropriately represented in the Visual Analytics solutions. The user needs to be aware of these data and analysis quality properties at any stage in the data analysis process.

Visual Representation and Level of Detail: To accommodate vast streams of data, appropriate solutions need to intelligently combine visualizations of selected analysis details on the one hand, and a global overview on the other hand. The relevant data patterns and relationships need to be visualized on several levels of detail, and with appropriate levels of data and visual abstraction.

User Interfaces, and Interaction Styles and Metaphors: Visual Analytics systems need to be easily used and interacted with by the analyst. The analyst needs to be able to fully focus on the task at hand, not on overly technical or complex user interfaces, which potentially distract. To this end, novel interaction techniques need to be developed which fully support the seamless, intuitive visual communication with the system. User feedback should be taken as intelligently as possible, requiring as little manual user input as possible, which guarantees the full support of the user in navigating and analyzing the data, memorizing insights and making informed decisions.

Display Devices: In addition to high-resolution desktop displays, advanced display devices such as large-scale power walls and small portable personal assistant, graphically-enabled devices need to be supported. Visual Analytics systems should adapt to the characteristics of the available output devices, supporting the Visual Analytics workflow on all levels of operation.

Evaluation: Due to the complex and heterogeneous problem domains addressed by Visual Analytics, so far it has been difficult to perform encompassing evaluation work. A theoretically founded evaluation framework needs to be developed which allows assessing the contribution of any Visual Analytics system toward the level of effectiveness and efficiency achieved regarding their requirements.

Infrastructure: Managing large amounts of data for visualization or analysis requires special data structures and mechanisms, both in memory and disks. Achieving interactivity means refreshing the display in 100ms at worst whereas analyzing data with standard techniques such as clustering can take hours to complete. Achieving the smooth interaction required by the analysts to perform their tasks while providing high-quality analytical algorithms need the combination of asynchronous computation with hybrid analytical algorithms that can trade time with quality. Moreover, to fully support the analytical process, the history of the analysis should also be recorded and interactively edited and annotated. Altogether, these requirements call for a novel software infrastructure, built upon well understood technologies such as databases, software components and visualization but augmented with asynchronous processing, history managements and annotations.

7   Examples for Visual Analytics Applications

7.1   Visual Analytics Tools for Analysis of Movement Data

With widespread availability of low cost GPS devices, it is becoming possible to record data about the movement of people and objects at a large scale. While these data hide important knowledge for the optimization of location and mobility oriented infrastructures and services, by themselves they lack the necessary semantic embedding which would make fully automatic algorithmic analysis possible. At the same time, making the semantic link is easy for humans who however cannot deal well with massive amounts of data. In [5] we argue that by using the right visual analytics tools for the analysis of massive collections of movement data, it is possible to effectively support human analysts in understanding movement behaviors and mobility patterns.

Figure 5 shows a subset of raw GPS measurements presented in so-called space-time cube. The large amount of position records referring to the same territory over a long time period makes it virtually impossible to do the analysis by purely visual methods.

The paper [5] proposes a framework where interactive visual interfaces are synergistically combined with database operations and computational processing. The generic database techniques are used for basic data processing and extraction of relevant objects and features. The computational techniques, which are specially devised for movement data, aggregate and summarize these objects and features and thereby enable the visualization of large amounts of information. The visualization enables human cognition and reasoning, which, in turn, direct and control the further analysis by means of the database, computational, and visual techniques. Interactive visual interfaces embrace all the tools.

Thus, in order to detect and interpret significant places visited by the moving entities, the positions of stops are extracted from the data by means of appropriate database queries. Then, clustering methods are applied to detect frequently visited places. Interactive visual displays put the results in the spatial and temporal contexts. The spatial positions of the stops can be observed on

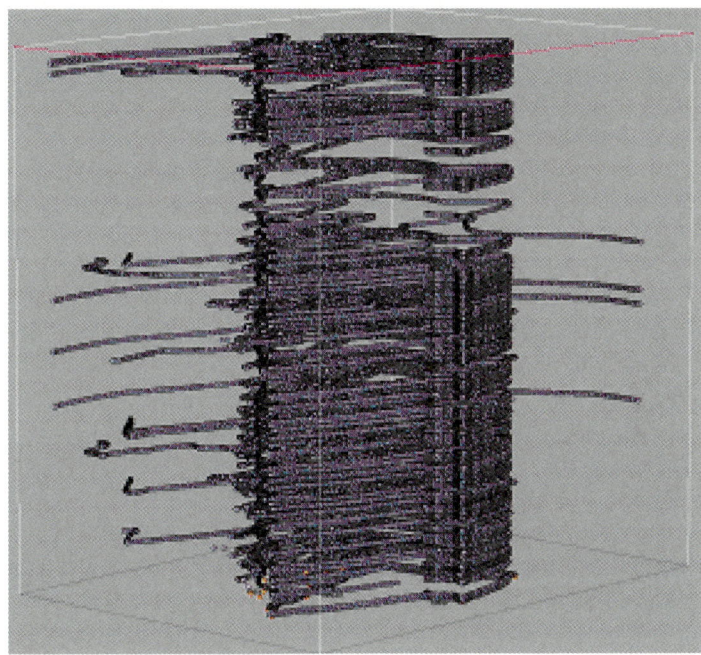

Fig. 5. A visual display of a large amount of position records is unreadable and not suitable for analysis.

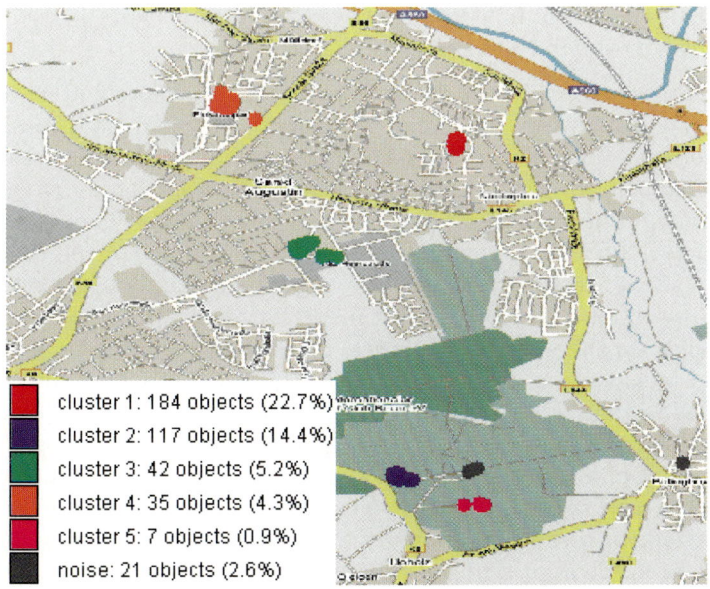

Fig. 6. Positions of stops have been extracted from the database. By means of clustering, frequently visited places have been detected.

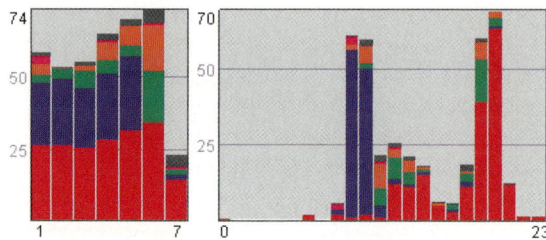

Fig. 7. The temporal histograms show the distribution of the stops in the frequently visited places (Figure 6) with respect to the weekly (left) and daily (right) cycles.

a map (Figure 6) or 3D spatial view. Temporal histograms (Figure 7) are used to explore the temporal distribution of the stops throughout the time period and within various temporal cycles (daily, weekly, etc.). These complementary views allow a human analyst to understand the meanings or roles of the frequently visited places.

In order to detect and interpret typical routes of the movement between the significant places, the analyst first applies a database query to extract sequences of position records between the stops, from which trajectories (time-referenced lines) are constructed. Then, clustering is applied with the use of specially devised similarity measures. The results are computationally generalized and summarized and displayed in the spatial context (Figure 8).

7.2   Multilevel Visualization of the Worldwide Air Transportation Network

The air transportation network has now become more dense and more complex at all geographical levels. Its dynamic no more rests on simple territorial logics. The challenge is to gain insightful understandings on how the routes carrying the densest traffic organize themselves and impact the organization of the network into sub-communities at lower levels. At the same time, subnetworks grow on their own logic, involving tourism, economy or territorial control, and influence or fight against each other. Because of the network size and complexity, its study can no more rely on traditional world map and requires novel visualization. A careful analysis of the network structural properties, requiring recent results on small world phenomenon, reveals its multilevel community structure.

The original network is organized into a top level network of communities (Figure 9(a)). Each component can then be further decomposed into subcommunities. Capitals such as New York, Chicago, Paris or London (Figure 9(b)) clearly attract most of the international traffic and impose routes to fly the world around because of airline partnerships (economical logic). Asia (Figure 9(c)) clearly stands apart from these core hubs because of strong territorial ties endorsed by national Asian airline companies (territorial logic). Visualization of social networks such as the worldwide air transportation is challenged by the necessity to scale with the growing size of network data while being able to offer

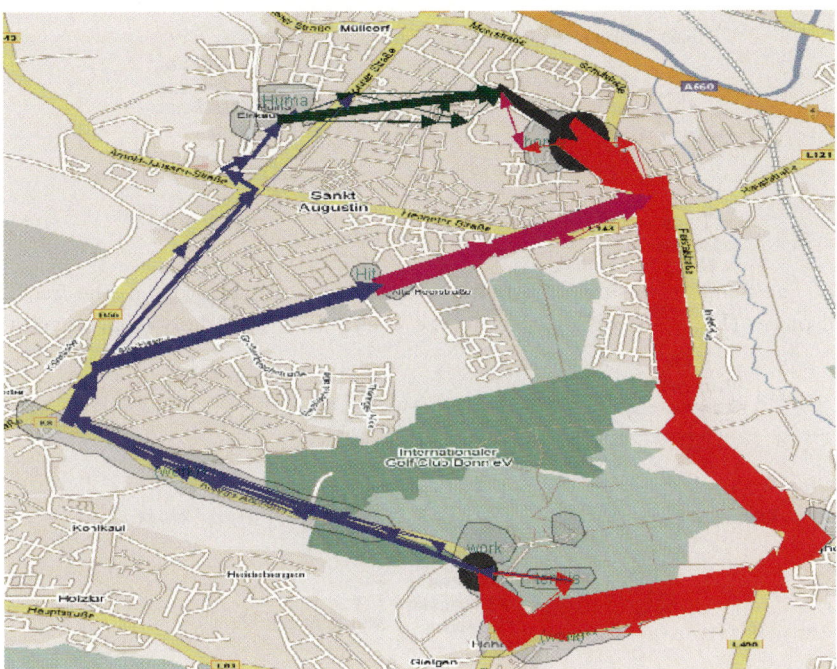

Fig. 8. A result of clustering and summarization of movement data: the routes between the significant places.

readable visual representations and fluid interaction. Visualization today brings the field of social sciences close to the study of complex systems and promises to deliver new knowledge across these disciplines [7,3,10].

8   Conclusions

The problems addressed by Visual Analytics are generic. Virtually all sciences and many industries rely on the ability to identify methods and models, which can turn data into reliable and provable knowledge. Ever since the dawn of modern science, researchers needed to find methodologies to create new hypotheses, to compare them with alternative hypotheses, and to validate their results. In a collaborative environment this process includes a large number of specialized people each having a different educational background. The ability to communicate results to peers will become crucial for scientific discourse.

Currently, no technological approach can claim to give answers to all three key questions that have been outlined in the first section, regarding the

- relevance of a specific information
- adequacy of data processing methods and validity of results
- acceptability of the presentation of results for a given task

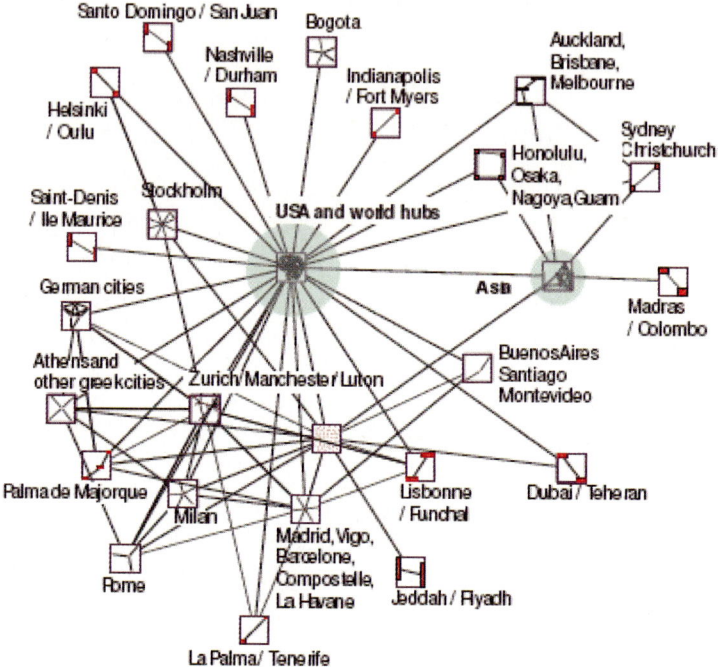

(a) World air transportation network.

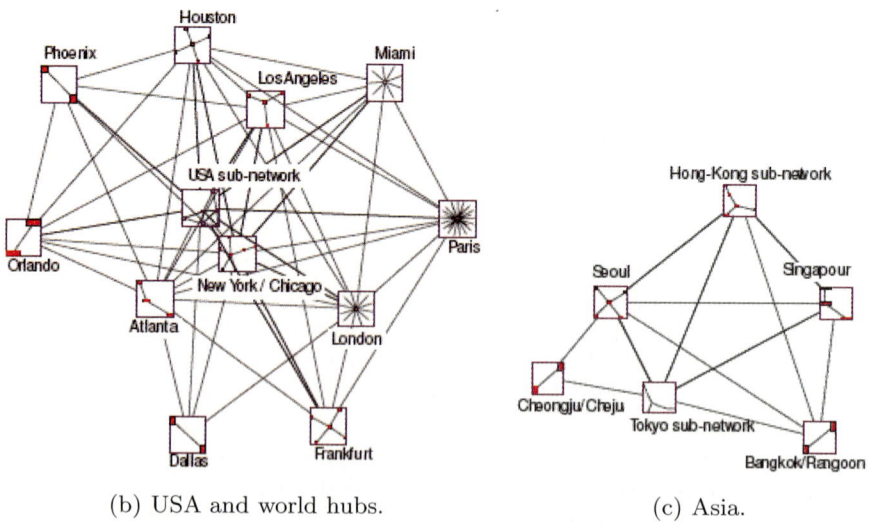

(b) USA and world hubs.   (c) Asia.

Fig. 9. Multilevel Visualization of the Worldwide Air Transportation Network

Visual Analytics research does not focus on specific methods to address these questions in a single "best-practice". Each specific domain contributes a repertoire of approaches to initiate an interdisciplinary creation of solutions.

Visual Analytics literally maps the connection between different alternative solutions, leaving the opportunity for the human user to view these options in the context of the complete knowledge generation process and to discuss these options with peers on common ground.

References

1. Aigner, W., Miksch, S., Müller, W., Schumann, H., Tominski, C.: Visual methods for analyzing time-oriented data. IEEE Transactions on Visualization and Computer Graphics 14(1), 47–60 (2008)
2. Amar, R.A., Eagan, J., Stasko, J.T.: Low-level components of analytic activity in information visualization. In: INFOVIS, p. 15 (2005)
3. Amiel, M., Melançon, G., Rozenblat, C.: Réseaux multi-niveaux: l'exemple des échanges aériens mondiaux. M@ppemonde 79(3) (2005)
4. Andrienko, G., Andrienko, N., Jankowski, P., Keim, D., Kraak, M.-J., MacEachren, A., Wrobel, S.: Geovisual analytics for spatial decision support: Setting the research agenda. Special issue of the International Journal of Geographical Information Science 21(8), 839–857 (2007)
5. Andrienko, G., Andrienko, N., Wrobel, S.: Visual analytics tools for analysis of movement data. ACM SIGKDD Explorations 9(2) (2007)
6. Andrienko, N., Andrienko, G.: Exploratory Analysis of Spatial and Temporal Data. Springer, Heidelberg (2005)
7. Auber, D., Chiricota, Y., Jourdan, F., Melançon, G.: Multiscale visualization of small world networks. In: INFOVIS (2003)
8. Card, S.K., Mackinlay, J., Shneiderman, B.: Readings in Information Visualization: Using Vision to Think. Morgan Kaufmann, San Francisco (1999)
9. Ceglar, A., Roddick, J.F., Calder, P.: Guiding knowledge discovery through interactive data mining, pp. 45–87. IGI Publishing, Hershey (2003)
10. Chiricota, Y., Melançon, G.: Visually mining relational data. International Review on Computers and Software (2005)
11. Das, A.: Semantic approximation of data stream joins. IEEE Transactions on Knowledge and Data Engineering 17(1), 44–59 (2005), Member-Johannes Gehrke and Member-Mirek Riedewald
12. Dix, A., Finlay, J.E., Abowd, G.D., Beale, R.: Human-Computer Interaction (.), 3rd edn. Prentice-Hall, Inc., Upper Saddle River (2003)
13. Duda, R., Hart, P., Stock, D.: Pattern Classification. John Wiley and Sons Inc., Chichester (2000)
14. Dykes, J., MacEachren, A., Kraak, M.-J.: Exploring geovisualization. Elsevier Science, Amsterdam (2005)
15. Engel, K., Hadwiger, M., Kniss, J.M., Rezk-salama, C., Weiskopf, D.: Real-time Volume Graphics. A. K. Peters, Ltd., Natick (2006)
16. Ester, M., Sander, J.: Knowledge Discovery in Databases - Techniken und Anwendungen. Springer, Heidelberg (2000)
17. Forsell, C., Seipel, S., Lind, M.: Simple 3d glyphs for spatial multivariate data. In: INFOVIS, p. 16 (2005)

18. Han, J., Kamber, M. (eds.): Data Mining: Concepts and Techniques. Morgan Kaufmann, San Francisco (2000)
19. Hand, D., Mannila, H., Smyth, P. (eds.): Principles of Data Mining. MIT Press, Cambridge (2001)
20. Inselberg, A., Dimsdale, B.: Parallel Coordinates: A Tool for Visualizing Multivariate Relations (chapter 9), pp. 199–233. Plenum Publishing Corporation, New York (1991)
21. Jacko, J.A., Sears, A.: The Handbook for Human Computer Interaction. Lawrence Erlbaum & Associates, Mahwah (2003)
22. Johnson, C., Hanson, C. (eds.): Visualization Handbook. Kolam Publishing (2004)
23. Keim, D., Ertl, T.: Scientific visualization (in german). Information Technology 46(3), 148–153 (2004)
24. Keim, D., Ward, M.: Visual Data Mining Techniques (chapter 11). Springer, New York (2003)
25. Keim, D.A., Ankerst, M., Kriegel, H.-P.: Recursive pattern: A technique for visualizing very large amounts of data. In: VIS '95: Proceedings of the 6th conference on Visualization '95, Washington, DC, USA, p. 279. IEEE Computer Society Press, Los Alamitos (1995)
26. Keim, D.A., Panse, C., Sips, M., North, S.C.: Pixel based visual data mining of geo-spatial data. Computers &Graphics 28(3), 327–344 (2004)
27. Kerren, A., Stasko, J.T., Fekete, J.-D., North, C.J. (eds.): Information Visualization. LNCS, vol. 4950. Springer, Heidelberg (2008)
28. Krúger, J., Schneider, J., Westermann, R.: Clearview: An interactive context preserving hotspot visualization technique. IEEE Transactions on Visualization and Computer Graphics 12(5), 941–948 (2006)
29. Maimon, O., Rokach, L. (eds.): The Data Mining and Knowledge Discovery Handbook. Springer, Heidelberg (2005)
30. Meliou, A., Chu, D., Guestrin, C., Hellerstein, J., Hong, W.: Data gathering tours in sensor networks. In: IPSN (2006)
31. Mitchell, T.M.: Machine Learning. McGraw-Hill, Berkeley (1997)
32. Naumann, F., Bilke, A., Bleiholder, J., Weis, M.: Data fusion in three steps: Resolving schema, tuple, and value inconsistencies. IEEE Data Eng. Bull. 29(2), 21–31 (2006)
33. North, C.: Toward measuring visualization insight. IEEE Comput. Graph. Appl. 26(3), 6–9 (2006)
34. Perner, P. (ed.): Data Mining on Multimedia Data. LNCS, vol. 2558. Springer, Heidelberg (2002)
35. Schumann, H., Müller, W.: Visualisierung - Grundlagen und allgemeine Methoden. Springer, Heidelberg (2000)
36. Shneiderman, B.: Tree visualization with tree-maps: 2-d space-filling approach. ACM Trans. Graph. 11(1), 92–99 (1992)
37. Shneiderman, B., Plaisant, C.: Designing the User Interface. Addison-Wesley, Reading (2004)
38. Spence, R.: Information Visualization. ACM Press, New York (2001)
39. Thomas, J.J., Cook, K.A.: Illuminating the Path. IEEE Computer Society Press, Los Alamitos (2005)
40. Tricoche, X., Scheuermann, G., Hagen, H.: Tensor topology tracking: A visualization method for time-dependent 2d symmetric tensor fields. Comput. Graph. Forum 20(3) (2001)
41. Unwin, A., Theus, M., Hofmann, H.: Graphics of Large Datasets: Visualizing a Million (Statistics and Computing). Springer, New York (2006)

42. van Wijk, J.J.: The value of visualization. In: IEEE Visualization, p. 11 (2005)
43. Widom, J.: Trio: A system for integrated management of data, accuracy, and lineage. In: CIDR, pp. 262–276 (2005)
44. Yi, J.S., Kang, Y.a., Stasko, J.T., Jacko, J.A.: Toward a deeper understanding of the role of interaction in information visualization. IEEE Trans. Vis. Comput. Graph. 13(6), 1224–1231 (2007)

Author Index

Printing: Mercedes-Druck, Berlin
Binding: Stein+Lehmann, Berlin

Lecture Notes in Computer Science

Sublibrary 3: Information Systems and Application, incl. Internet/Web and HCI

For information about Vols. 1– 4721
please contact your bookseller or Springer

Vol. 4903: S. Satoh, F. Nack, M. Etoh (Eds.), Advances in Multimedia Modeling. XIX, 510 pages. 2008.

Vol. 4900: S. Spaccapietra (Ed.), Journal on Data Semantics X. XIII, 265 pages. 2008.

Vol. 4892: A. Popescu-Belis, S. Renals, H. Bourlard (Eds.), Machine Learning for Multimodal Interaction. XI, 308 pages. 2008.

Vol. 4882: T. Janowski, H. Mohanty (Eds.), Distributed Computing and Internet Technology. XIII, 346 pages. 2007.

Vol. 4881: H. Yin, P. Tino, E. Corchado, W. Byrne, X. Yao (Eds.), Intelligent Data Engineering and Automated Learning - IDEAL 2007. XX, 1174 pages. 2007.

Vol. 4877: C. Thanos, F. Borri, L. Candela (Eds.), Digital Libraries: Research and Development. XII, 350 pages. 2007.

Vol. 4872: D. Mery, L. Rueda (Eds.), Advances in Image and Video Technology. XXI, 961 pages. 2007.

Vol. 4871: M. Cavazza, S. Donikian (Eds.), Virtual Storytelling. XIII, 219 pages. 2007.

Vol. 4858: X. Deng, F.C. Graham (Eds.), Internet and Network Economics. XVI, 598 pages. 2007.

Vol. 4857: J.M. Ware, G.E. Taylor (Eds.), Web and Wireless Geographical Information Systems. XI, 293 pages. 2007.

Vol. 4853: F. Fonseca, M.A. Rodríguez, S. Levashkin (Eds.), GeoSpatial Semantics. X, 289 pages. 2007.

Vol. 4836: H. Ichikawa, W.-D. Cho, I. Satoh, H.Y. Youn (Eds.), Ubiquitous Computing Systems. XIII, 307 pages. 2007.

Vol. 4832: M. Weske, M.-S. Hacid, C. Godart (Eds.), Web Information Systems Engineering – WISE 2007 Workshops. XV, 518 pages. 2007.

Vol. 4831: B. Benatallah, F. Casati, D. Georgakopoulos, C. Bartolini, W. Sadiq, C. Godart (Eds.), Web Information Systems Engineering – WISE 2007. XVI, 675 pages. 2007.

Vol. 4825: K. Aberer, K.-S. Choi, N. Noy, D. Allemang, K.-I. Lee, L. Nixon, J. Golbeck, P. Mika, D. Maynard, R. Mizoguchi, G. Schreiber, P. Cudré-Mauroux (Eds.), The Semantic Web. XXVII, 973 pages. 2007.

Vol. 4823: H. Leung, F. Li, R. Lau, Q. Li (Eds.), Advances in Web Based Learning – ICWL 2007. XIV, 654 pages. 2008.

Vol. 4822: D.H.-L. Goh, T.H. Cao, I.T. Sølvberg, E. Rasmussen (Eds.), Asian Digital Libraries. XVII, 519 pages. 2007.

Vol. 4820: T.G. Wyeld, S. Kenderdine, M. Docherty (Eds.), Virtual Systems and Multimedia. XII, 215 pages. 2008.

Vol. 4816: B. Falcidieno, M. Spagnuolo, Y. Avrithis, I. Kompatsiaris, P. Buitelaar (Eds.), Semantic Multimedia. XII, 306 pages. 2007.

Vol. 4813: I. Oakley, S.A. Brewster (Eds.), Haptic and Audio Interaction Design. XIV, 145 pages. 2007.

Vol. 4810: H.H.-S. Ip, O.C. Au, H. Leung, M.-T. Sun, W.-Y. Ma, S.-M. Hu (Eds.), Advances in Multimedia Information Processing – PCM 2007. XXI, 834 pages. 2007.

Vol. 4809: M.K. Denko, C.-s. Shih, K.-C. Li, S.-L. Tsao, Q.-A. Zeng, S.H. Park, Y.-B. Ko, S.-H. Hung, J.-H. Park (Eds.), Emerging Directions in Embedded and Ubiquitous Computing. XXXV, 823 pages. 2007.

Vol. 4808: T.-W. Kuo, E. Sha, M. Guo, L.T. Yang, Z. Shao (Eds.), Embedded and Ubiquitous Computing. XXI, 769 pages. 2007.

Vol. 4806: R. Meersman, Z. Tari, P. Herrero (Eds.), On the Move to Meaningful Internet Systems 2007: OTM 2007 Workshops, Part II. XXXIV, 611 pages. 2007.

Vol. 4805: R. Meersman, Z. Tari, P. Herrero (Eds.), On the Move to Meaningful Internet Systems 2007: OTM 2007 Workshops, Part I. XXXIV, 757 pages. 2007.

Vol. 4804: R. Meersman, Z. Tari (Eds.), On the Move to Meaningful Internet Systems 2007: CoopIS, DOA, ODBASE, GADA, and IS, Part II. XXIX, 683 pages. 2007.

Vol. 4803: R. Meersman, Z. Tari (Eds.), On the Move to Meaningful Internet Systems 2007: CoopIS, DOA, ODBASE, GADA, and IS, Part I. XXIX, 1173 pages. 2007.

Vol. 4802: J.-L. Hainaut, E.A. Rundensteiner, M. Kirchberg, M. Bertolotto, M. Brochhausen, Y.-P.P. Chen, S.S.-S. Cherfi, M. Doerr, H. Han, S. Hartmann, J. Parsons, G. Poels, C. Rolland, J. Trujillo, E. Yu, E. Zimányie (Eds.), Advances in Conceptual Modeling – Foundations and Applications. XIX, 420 pages. 2007.

Vol. 4801: C. Parent, K.-D. Schewe, V.C. Storey, B. Thalheim (Eds.), Conceptual Modeling - ER 2007. XVI, 616 pages. 2007.

Vol. 4797: M. Arenas, M.I. Schwartzbach (Eds.), Database Programming Languages. VIII, 261 pages. 2007.

Vol. 4796: M. Lew, N. Sebe, T.S. Huang, E.M. Bakker (Eds.), Human–Computer Interaction. X, 157 pages. 2007.

Vol. 4794: B. Schiele, A.K. Dey, H. Gellersen, B. de Ruyter, M. Tscheligi, R. Wichert, E. Aarts, A. Buchmann (Eds.), Ambient Intelligence. XV, 375 pages. 2007.

Vol. 4777: S. Bhalla (Ed.), Databases in Networked Information Systems. X, 329 pages. 2007.

Vol. 4761: R. Obermaisser, Y. Nah, P. Puschner, F.J. Rammig (Eds.), Software Technologies for Embedded and Ubiquitous Systems. XIV, 563 pages. 2007.

Vol. 4747: S. Džeroski, J. Struyf (Eds.), Knowledge Discovery in Inductive Databases. X, 301 pages. 2007.

Vol. 4744: Y. de Kort, W. IJsselsteijn, C. Midden, B. Eggen, B.J. Fogg (Eds.), Persuasive Technology. XIV, 316 pages. 2007.

Vol. 4740: L. Ma, M. Rauterberg, R. Nakatsu (Eds.), Entertainment Computing – ICEC 2007. XXX, 480 pages. 2007.

Vol. 4730: C. Peters, P. Clough, F.C. Gey, J. Karlgren, B. Magnini, D.W. Oard, M. de Rijke, M. Stempfhuber (Eds.), Evaluation of Multilingual and Multi-modal Information Retrieval. XXIV, 998 pages. 2007.

Vol. 4723: M. R. Berthold, J. Shawe-Taylor, N. Lavrač (Eds.), Advances in Intelligent Data Analysis VII. XIV, 380 pages. 2007.